3000 800022 6936
St. Louis Community College

C0-AQW-653

WITHDRAWN

FV

 St. Louis Community College

Forest Park
Florissant Valley
Meramec

Instructional Resources
St. Louis, Missouri

DESIGN
OF REINFORCED
CONCRETE
STRUCTURES

DESIGN OF REINFORCED CONCRETE STRUCTURES

SECOND EDITION

HENRY J. COWAN

University of Sydney
Australia

PRENTICE HALL, ENGLEWOOD CLIFFS, NEW JERSEY 07632

Library of Congress Cataloging-in-Publication Data

COWAN, HENRY J.
    Design of reinforced concrete structures / Henry J. Cowan—2nd
    ed.
        p.   cm.
    Includes bibliographies and index.
    ISBN   0-13-201443-2
    1. Reinforced concrete construction.   2. Structural design.
    I. Title.
    TA683.2.C63   1989   87-35202
    624.1'8341—dc19   CIP

Editorial/production supervision and
    interior design: Ellen Greenberg & Debbie Young
Cover design: Diane Saxe
Cover photograph: Ezra Stoller © *ESTO. All rights reserved.*
Manufacturing buyer: Mary Noonan

© 1989 by Prentice-Hall, Inc.
A Division of Simon & Schuster
Englewood Cliffs, New Jersey 07632

All rights reserved. No part of this book may be
reproduced, in any form or by any means,
without permission in writing from the publisher.

Printed in the United States of America

10 9 8 7 6 5 4 3 2 1

ISBN 0-13-201443-2

Prentice-Hall International (UK) Limited, *London*
Prentice-Hall of Australia Pty. Limited, *Sydney*
Prentice-Hall Canada Inc., *Toronto*
Prentice-Hall Hispanoamericana, S.A., *Mexico*
Prentice-Hall of India Private Limited, *New Delhi*
Prentice-Hall of Japan, Inc., *Tokyo*
Simon & Schuster Asia Pte. Ltd., *Singapore*
Editora Prentice-Hall do Brasil, Ltda., *Rio de Janeiro*

To William Cohn, Dallas

Contents

**CHAPTER 9   ULTIMATE-STRENGTH DESIGN OF REINFORCED
CONCRETE SLABS AND RECTANGULAR BEAMS   99**

**CHAPTER 10 THE MONOLITHIC CONSTRUCTION OF BEAMS
AND SLABS   133**

CHAPTER 11 SHEAR AND TORSION   153

Preface

This book explains the principles of the structural design of reinforced concrete buildings without going into all the details of a very complex subject. In addition, it explains fully the structural design of reinforced concrete slabs, which frequently occur in small buildings that do not necessarily call for the services of a structural engineer. The text is particularly intended to meet the needs of architecture and construction students. They require a general knowledge of reinforced concrete design, without having to master all the intricacies of the concrete code. They may also wish to design a small reinforced concrete floor or roof in a building that otherwise does not require structural calculations.

The first three chapters provide a general introduction to the use of reinforced concrete in buildings. Chapters 4 and 5 describe the concrete code, the systems of measurement, and the properties of the materials. Chapter 6 explains the detailing of the reinforcement that is essential for design and construction. Chapters 7 and 8 discuss the procedure used in reinforced concrete design and the assumptions on which it is based.

In Chapter 9 the theory of reinforced concrete slabs and rectangular beams is derived, and several slabs and beams are fully designed and detailed. Chapters 10 and 11 deal briefly with T-beams, compression reinforcement, shear, and torsion, including several design examples. In Chapter 12 panels of two-way slabs and flat plates are designed. Chapters 13 and 14 deal with columns and footings, again only briefly but with the aid of some examples. When the reader has mastered this subject matter, he or she should be able to design and detail a very simple reinforced concrete structure, and understand how more complicated structures are designed, even though not able to do it himself or herself.

Chapters 15 and 16 deal with two related topics, prestressed concrete, and concrete shells and folded plates, including a few simple examples.

I am indebted to Sydney University Press for permission to use a part of the third edition of a similar book written in conformity with the Australian Concrete Code, but the text has been changed to conform in every respect to the 1983 ACI Code.

I should like to acknowledge the courtesy with which the American Concrete Institute, the American Society for Testing and Materials, the Prestressed Concrete Institute, and the Canadian Standards Association dealt with my long-distance enquiries. I should like to express my appreciation to Dr. S. Aroni and Dr. C. P. Siess for reading the manuscript.

HENRY J. COWAN

Chapter 1

Introduction

This chapter describes briefly what concrete is, and why and where we reinforce it.   In explains what we mean by cast-in-place concrete and by monolithic construction, and it compares reinforced concrete and steel as structural materials.

If you already know what reinforced concrete is, you need not read this chapter.

1.1 CONCRETE AS A MATERIAL

Normal concrete is an artificial stone made from crushed rock (or crushed or whole gravel), sand, cement, and water.   The materials are mixed together until a dense, uniform, and plastic mix is obtained.   The mixture is then placed in a mold and allowed to set and harden.

The resultant material resembles natural stone in many of its properties. It is hard and brittle, strong in compression and weak in tension; its weight ranges from 100 to 150 pcf (1 500 to 2 500 kg/m^3) and its modulus of elasticity is from 3,000,000 to 4,000,000 psi (20 000 to 30 000 MPa).

Unlike natural stone, it can be produced in any shape without having to resort to cutting tools.   The whole structure can be cast in one piece, and the expense of joining individual blocks to one another and the weakness resulting from the presence of joints is thus eliminated.

The surface finish of concrete requires careful attention.   Whereas most natural stones grow old gracefully without special care, concrete does so only if its finish has been properly designed.

Concrete, like timber, deforms as its moisture content changes, and its deflection increases with time.   In this respect it compares unfavorably with both natural stone and steel.

On the other hand, concrete is not affected by any but the most severe fires in buildings, and in this respect it is superior to both steel and timber.

1.2 NORMAL AND LIGHTWEIGHT CONCRETE

Most concrete is made with gravel or crushed rock as the coarse (or largest-size) aggregate and this determines its weight.   The biggest load carried by the structure is usually its own weight, and there are occasions when it is economical to use a lighter aggregate even though it costs more.

Lightweight concrete is a concrete made with a lightweight aggregate (see Sections 2.1 and 5.3), and weighing about 100 pcf (1 500 kg/m^3).   Much lighter concrete can be produced by eliminating the aggregate altogether, and introducing air bubbles into the concrete.   However, most of these very light concretes are suitable only for insulation, not for structural concrete, and they are beyond the scope of this book.

1.3 PLAIN, REINFORCED, AND PRESTRESSED CONCRETE

The scope of traditional masonry construction is severely limited by the relative weakness of brick and natural stone in tension, and by the need for a larger number of joints with even lower tensile strength.   Since concrete is cast in one piece, or at least in a few very large pieces, the weakness of the mortar joints is eliminated.   However, the limitation due to the low tensile strength of concrete (which varies from $\frac{1}{10}$ to $\frac{1}{20}$ of its compressive strength) remains in plain concrete construction.   Hence, plain concrete is used only for footings, concrete slabs laid on the ground, and for massive structures such as retaining walls; even then reinforcement is frequently employed.

This disability of concrete can be overcome in two ways, both requiring the use of steel.   Steel bars can be cast into the concrete, so as to take the tensile component of the bending moment (Fig. 1.1).   They do not prevent the concrete from cracking, but if the cracks are kept very small and are bridged by tension steel, they have no adverse effect on the safety or durability of the structure.

The presence of fine cracks in reinforced concrete is inevitable.   The stress in the lowest-grade reinforcing steel under the normal working loads is of the order of 20,000 psi (140 MPa) (See Section 8.1.). Taking the modulus of elasticity of steel as 29,000,000 psi (200 000 MPa), this amounts to an elastic

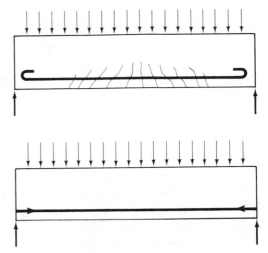

Figure 1.1 Reinforced concrete cracks
under working loads. Prestressed
concrete cracks only under an overload.

strain of 7×10^{-4}, which is more than the ultimate tensile strain of concrete.
Cracks are thus produced in the concrete by the mere process of the reinforcing steel being stressed under the normal load.   It is perhaps fortunate
that this was not understood when reinforced concrete was first employed
more than a century ago; otherwise, building authorities with a reasonable
concern for public safety would probably have forbidden its use.

The safety of reinforced concrete structures depends on the width of
the cracks being kept below a permissible minimum (see Section 6.2), and
this has become a more serious problem in recent years because the use of
higher steel stresses also increases the strain of the concrete.   At the same
time the use of higher-strength concrete does not greatly increase its ultimate
tensile strain.   Thus crack control is a more serious matter in reinforced
concrete design than it was, say, 20 years ago.

Alternatively, an initial compression can be introduced into the concrete
by placing *prestressed* steel in the regions where tension is expected, thus
inhibiting the formation of cracks.   In this respect, prestressed concrete is
superior to reinforced concrete, but it has other limitations.   Prestressed
concrete is discussed briefly in Chapter 15.   The discussion in this book,
however, deals primarily with normal reinforced concrete in which steel bars
are cast into the concrete to provide additional strength and bridge any cracks
that may form.

1.4 PRECAST AND CAST-IN-PLACE CONCRETE

A reinforced concrete structure may be built up from a number of component
parts, cast in separate molds (i.e., precast), and then assembled into a structure.   Alternatively, the complete structure may be cast in place (i.e., on the

building site in its actual position), and this is the normal procedure. Unless reinforced concrete is specifically designated as precast, it is meant to be cast in place.

Precast concrete may be made in a factory and then transported to the site after the concrete has set and has gained sufficient strength to be handled without damage. It is easier to produce a concrete of good quality under factory conditions, and in particular a good surface finish can be readily produced when a component is cast horizontally. Precasting is therefore particularly useful for concrete panels to be used in walls, for which a good surface finish would be much harder to produce in its final vertical position.

Concrete can also be precast on the site. This avoids transportation by road, as it is necessary only to lift the component into position, but it necessitates setting up a casting yard on the site, which may have less stringent control over concrete quality.

Precasting was used extensively in the late nineteenth century because the theory of reinforced concrete design had not been fully developed. Precast units could be tested to destruction, which gave a satisfactory assurance of the strength of the other identical components. There was another upsurge in the use of precasting in the 1930s and 1940s, when cast-in-place concrete was still mixed in a concrete mixer on the building site. The mixer produced a great deal of dust and occupied valuable space. The balance was tilted again in favor of cast-in-place concrete by the development of ready-mixed concrete, which removed the mixing operation to a factory while leaving the casting operation on the site.

The relative advantages of cast-in-place and precast concrete depend to a large extent on the climate and on the number of units required. Concrete can be cast in a factory throughout the year, but in a cold climate cast-in-place construction is inhibited, or at least made much more difficult in winter because the chemical reaction in the cement is slowed down and eventually stopped by low temperatures. The capital cost of a factory and of transport specially fitted for large concrete units is spread over a great number of units as the production runs are increased.

The removal of the mixing operation to the ready-mix factory leaves one major disadvantage of cast-in-place concrete—the risk of spilling wet concrete on surfaces already installed. On the other hand, the transport of precast units, and their hoisting into position on the site, may also cause damage, notably to the units themselves, for example, broken edges and corners.

At the present time most structural reinforced concrete is cast in place. Usually, this requires a number of casting operations, but if proper construction joints are used, then the columns, the beams, and the floor slabs (and occasionally the walls) act *monolithically* (i.e., like a single piece of stone), and the framework has great rigidity. The rigidity of the monolithic frame, and the ability to cast it in its final position from cheap and universally

available raw materials, are two of the most important advantages of reinforced concrete construction.

1.5 MONOLITHIC BEHAVIOR OF THE CAST-IN-PLACE REINFORCED CONCRETE FRAME

Monolithy (i.e., behavior as if cut from one single piece of stone) can be
established in precast concrete construction by making rigid connections between the individual members, although this is not normally considered necessary.   In structural steel, also, rigidity can be achieved by the use of stiff
connections using welding or high-tensile bolting.   Concrete, however, is
most easily cast in one piece.   The structure is a continuous frame in which
all members are rigidly connected to one another.

A load placed on such a frame affects the entire structure.   The effect
of the load is most pronounced in the loaded member, is of appreciable
magnitude in the adjacent member, and gradually decreases to negligible
proportions in the more distant members.   The load is therefore supported
not only by the structural member on which it is acting, but also by the
members to which it is connected.

This leads to an economy of material, because the bending moments
are reduced, and the size of the members may be decreased accordingly.   It
also complicates the design of the structural members, although this is of
much less significance since the introduction of computerized frame analysis.

It is incorrect to treat a structural member of a cast-in-place reinforced
concrete structure as if it were simply supported.   Generally, a rigid frame
has negative bending moments over the supports of beams and slabs, and
these require reinforcements on top, where the tension occurs; at midspan
the reinforcement is required at the bottom.   In structural steel and in timber
construction a mistake in the sign of the bending moment is not very serious,
because the behavior of these materials in tension and compression is approximately the same.   In reinforced concrete, placing the tension steel on
the wrong face could lead to collapse of the structure.

1.6 FUNCTION OF STEEL IN REINFORCED CONCRETE

Although steel occupies only a small part of the volume of reinforced concrete
(on the average about 1%), it is a major part of the cost.   As a very rough
guide, the cost of the formwork, the cost of the concrete, and the cost of the
steel are approximately the same, that is, one-third of the total.

The steel is also the vital part of the structure, since concrete is deficient
in tensile strength.   It is possible to build a structural frame from steel without
concrete, but not from concrete without steel.

Tension is not merely caused by the action of the loads, but also by the shrinkage and temperature movement of the concrete (see Section 8.2). It is therefore necessary to provide a small amount of reinforcement on all faces of major structural members, even if conventional calculations for stresses due to the loads do not indicate the need for doing so. A section at which, for any reason, tensile stresses occur on a face without reinforcement has no more strength than a similar plain concrete section.

Pier Luigi Nervi (Ref. 1.1) developed a material that he called *ferro-cemento*, consisting of layers of fine wire mesh filled and surrounded by sand-cement mortar. The reinforcement is thus finely distributed throughout the body of the mortar, and this gives great resilience to the resulting structure, but also increases the cost of labor. This material falls outside the scope of this book, but it illustrates the importance of ensuring the proper integration of the reinforcing steel and the concrete.

Since reinforcing steel rusts very readily, it must be protected by an adequate concrete cover to ensure the durability of the structure (see Section 6.1). Galvanizing of reinforcing steel as a protection from rust is still an exceptional measure. The steel is therefore completely surrounded by the concrete, and the loads act on the concrete; thus the strain can be transmitted to the steel only via the concrete. The proper adhesion between the steel and the concrete is of the greatest importance, and bars should be of sufficiently small diameter to offer an adequate area of contact with the steel; note that the smaller the diameter of the bars, the greater their surface area for any given percentage of reinforcement. The practical limit is reached when the bars become so numerous that they obstruct the proper placing of concrete.

The integration of the steel and the concrete requires the provision of adequate concrete around each reinforcing bar to ensure proper bond between the steel and the concrete. As a result, there are rules for minimum spaces between bars (see Section 6.4).

Although the structure is made in one piece, the steel reinforcing bars are not. The component pieces of a rigid structural steel frame are joined together by welding or other means, and the same could be done to the bars in a concrete structure to produce a rigid form of reinforcement. This would, however, be a considerable and unnecessary expense. We are therefore dealing with individual reinforcing bars which generally do not exceed 60 ft (18 m) in length because longer bars are difficult to transport. These bars are joined by bonding them to the concrete. The steel stress is thus transmitted to the concrete by bond or anchorage, and it is then transmitted to another bar by the same means. It is thus necessary to provide adequate space for the reinforcement to develop the stress by transmission from the concrete (see Section 6.7).

At the junction of several structural members, their reinforcing bars must pass one another (Fig. 1.2); it is not economical to cut the bars and

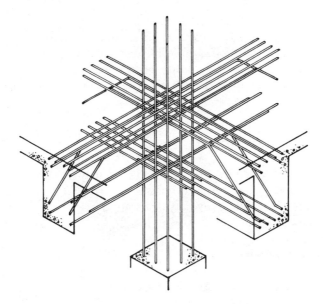

Figure 1.2   Typical reinforcement detail at the junction of two main beams and a column.

reweld them as is done in structural-steel frames.   The detailing of the reinforcement at a junction thus requires particular attention.

1.7 COMPARATIVE ADVANTAGE OF STEEL AND REINFORCED CONCRETE

Fifty years ago structural steel was the normal structural material for all but the smallest building frames.   Today the economic balance has tilted in favor of reinforced concrete.   This is part of a worldwide trend, although it is more pronounced in Europe than in North America.

Since the choice between the two materials is a fundamental one that must be made at the beginning of the design process, we will examine the principal arguments determining the choice.

Control over the Quality of the Materials

Steel is produced and fabricated under factory conditions, and erection is carried out by skilled labor specializing in this form of construction.

Cement is a mass-produced, relatively cheap material with more variation from batch to batch than would be acceptable in steel.   Aggregate and sand are materials dug from the ground, which makes rigid quality control impracticable.   The mixing of concrete in a ready-mix plant is generally well controlled, but placing of the concrete on the site is more variable. The rigid rules for quality control of concrete are designed to ensure that the concrete has the specified strength. They are, however, contingent on proper supervision and testing (see Section 5.8). Moreover, the result of the tests is obtained only after the concrete has hardened, and corrective measures are then difficult.

Availability of Materials

Reinforced concrete uses less steel than structural steel (but more than prestressed concrete), and at times of a steel shortage it is often used for that reason. Cement factories are widely distributed around the world, and concrete aggregates are generally available locally.

Speed of Erection

It is necessary for concrete columns to gain sufficient strength by chemical changes in the cement before it is possible to put a load on them, and the concrete floor must also be left for a few days before it is possible to work on it (the floor structure can be left supported by the floor below while work proceeds).

Structural steel arrives on the site capable of taking its full stress, and erection is less likely to be affected by the weather. A large part of the fabrication has been done in the factory, and erection is therefore much quicker.

On the other hand, the structural steel frame does not provide working platforms as useful as those of a reinforced concrete frame (which invariably includes the floor) and more work remains to be done before the building is complete.

There is often little difference between the completion date of a steel-framed and a concrete-framed *building*, in spite of the fast erection of the steel *frame*.

Durability, Maintenance, and Fire Resistance

Reinforced concrete is fireproof, provided that there is adequate cover over the reinforcing steel. In any case, the cover required to protect the reinforcement against corrosion is often sufficient, although a high fire rating may require a small increase.

Structural steel is not fireproof, and the steel frame of multistory buildings consequently requires additional material for fireproofing. If the steel frame is encased in concrete for fire protection, the resulting frame sometimes contains almost as much concrete and appreciably more steel than does a reinforced concrete frame with the same load-bearing capacity. In North

America lightweight fireproofing with materials such as sprayed vermiculite is commonly used. A recent innovation is the cooling of hollow structural steel sections with water circulating through the structural frame, and conveying the heat in the steel to other parts of the building not affected by the fire.

Single-story structures generally do not require fireproofing, and structural steel then has an advantage. However, a bare steel frame needs to be protected against corrosion, and some protective coatings, particularly paint, require periodic maintenance. Concrete, on the other hand, is a durable material that normally requires no preventive maintenance during the entire life of the building.

Floor Space Occupied by the Columns

Because the strength of concrete is much lower than the strength of steel, reinforced concrete columns for tall buildings occupy more ground space than structural steel columns carrying a similar load and thus reduce the space available to the owner of the building. This has in the past been regarded as a severe limiting factor for the use of structural concrete in tall buildings, since column size inevitably increases with height (see Section 13.1). This can be partly overcome by the use of lightweight concrete, which has almost the same strength as normal concrete but, being lighter, reduces the weight of the structure. A more important advance is the introduction of new structural systems that eliminate interior columns and rely on exterior columns placed partly or wholly outside the building.

It is noteworthy that the majority of the tallest buildings in every country outside North and South America and Japan have a reinforced concrete structure.

Integration of the Body of the Building with the Structural Frame

A reinforced concrete frame usually includes the roof, the floors, and occasionally some of the vertical walls. The horizontal and vertical slabs contribute greatly to the strength of the frame, and allowance can be made for this contribution in the design. A structural steel frame is normally only a skeleton. The roof, the floors, and the walls constitute a dead load to be carried by the frame.

It should be noted that even in steel-framed buildings, the services core (containing the elevators) is frequently made of reinforced concrete.

Reinforced concrete has a particular advantage for long-span single-cell buildings, such as covered sporting areas, auditoria, and exhibition halls, which consist essentially of a floor, a roof, and banked seats, because the concrete roof and banked seating can be used both for their functional purpose and for the structure.

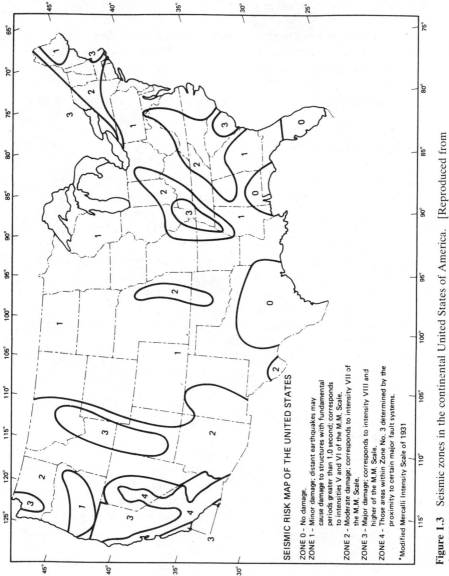

SEISMIC RISK MAP OF THE UNITED STATES

ZONE 0 – No damage.
ZONE 1 – Minor damage; distant earthquakes may
cause damage to structures with fundamental
periods greater than 1.0 second; corresponds
to intensities V and VI of the M.M. Scale.
ZONE 2 – Moderate damage; corresponds to intensity VII of
the M.M. Scale.
ZONE 3 – Major damage; corresponds to intensity VIII and
higher of the M.M. Scale.
ZONE 4 – Those areas within Zone No. 3 determined by the
proximity to certain major fault systems.

*Modified Mercalli Intensity Scale of 1931

Figure 1.3 Seismic zones in the continental United States of America. [Reproduced from the 1970 Uniform Building Code (Ref. 1.3); copyright 1979, The International Conference of Building Officials.]

1.8 EARTHQUAKE LOADING

An earthquake is a rapid and jerky movement of the ground, which takes the foundation of the building with it but leaves the upper part of the building behind because of the high speed of the ground's motion and the high inertia of the building.   The effect is the same as if the building moved relative to the ground.   The frame of the building must be sufficiently ductile to absorb the energy and dissipate it through its motion.   In earthquake zones the reinforcement must be designed to ensure that its ductility can be brought into play (Ref. 1.2), because the concrete is a brittle (i.e., nonductile) material (see Sections 9.3, 10.7, 11.6, and 13.11).

The main earthquake zones are the Circum-Pacific Belt, which contains Chile, Central America, California, Alaska, Japan, Indonesia, and New Zealand; and the Alpide Belt, which runs through Iran, Turkey, Greece, Yugoslavia, Italy, and Portugal.   The earthquake zones in the continental United States are shown in Fig. 1.3.

REFERENCES

1.1. P. L. Nervi, *Structures*, trans. M. and G. Salavadori, Dodge, New York, 1956.
1.2. ACI Committee 315, "Seismic Details for Special Ductile Frames," *Journal of the American Concrete Institute*, Vol. 67, 1970, p. 374.
1.3. *Uniform Building Code*, International Conference of Building Officials, Pasadena, Calif., 1979, pp. 145–146.

Chapter 2

The History
of Reinforced Concrete

This chapter is a brief account of the historical background of reinforced concrete. Both concrete and reinforcement have a prehistory, but reinforced concrete is a development of the second part of the nineteenth century, and rules for its design were first made at the turn of this century.

2.1 ROMAN CONCRETE

Sands of volcanic origin with cementing properties occur in many parts of the Mediterranean. When mixed with lime they form a mortar with properties not unlike those of modern cement mortar. One notable example is the Greek island of Thera (also known as Santorini, a name that it received during the Fourth Crusade), where the local volcanic soil is still used as a building material (Ref. 2.1).

In Imperial Rome concrete construction reached a degree of sophistication that it did not regain until the nineteenth century. This was probably due to a combination of two circumstances: the great skill of the Romans in construction, and the ready availability near Rome of volcanic sand with cementing properties. This was used (1) for *opus signinum*, a mortar with broken potsherds or broken bricks, and (2) *opus caementitium*, a mortar cast around large pieces of natural stone, or rubble from the demolition of previous buildings, used for enormous structures which have proved very durable.

For particularly important work, *pulvis puteolanus* was imported from the Roman port of Puteoli (the modern Pozzuoli, near Naples). The material has become world-famous under the name *pozzolana*, a term still used for certain cements (although with different properties). Vitruvius, who wrote in the first century B.C., describes it as follows:

> There is also a kind of powder which from natural causes produces astonishing results. It is found in the neighborhood of Baiae and in the country belonging to the towns around Mount Vesuvius. This substance, when mixed with lime and rubble, not only lends strength to buildings of other kinds, but even when piers of it are constructed in the sea, they set hard under water. The reason for this seems to be that the soil in the slopes of the mountains in these neigh-bourhoods is hot and full of hot springs. This would not be so unless the mountains had beneath them huge fires of burning sulfur or alum or asphalt. So the fire and the heat of the flames, coming up hot from far within through the fissures, make the soil light, and the tufa found is spongy and free from moisture. Hence the three substances, all formed on a similar principle by the force of fire, are mixed together, the water taken in makes them cohere, and the moisture quickly hardens them so that they can set into a mass which neither the waves nor the force of water can dissolve. (Ref. 2.2)

Vitruvius's explanation of the behavior of a cementing substance in terms of the classical Four Elements does not accord with the findings of modern chemistry. It does, however, draw attention to an important distinction between lime mortar, used throughout the world until well into the twentieth century, and cement mortar. Lime mortar is water-soluble, and is thus washed out by water. A "hydraulic mortar," made with *pulvis puteolanus* and lime, or with modern portland cement, is waterproof and can thus be used under water for hydraulic works.

Somewhat confusingly for the modern reader, the Latin word for concrete aggregate is *caementum*. *Opus caementitium* is thus work that has *caementa*, or large pieces of aggregate, embedded in the mortar. Roman aggregate varied greatly in its composition and size. For unimportant work, broken stone or concrete obtained from the demolition of old buildings was used, but for important buildings the aggregate was very carefully selected. The heavier materials were used at the bottom, and especially lightweight aggregates, such as pumice, were used higher up. Pieces were generally much larger than in modern concrete, and placed in position in regular layers before the mortar was poured in between.

The Pantheon (Fig. 2.1) had a span of 143 ft (44 m), the longest span prior to the nineteenth century, and this made the relative weight of the *caementa* very important (Ref. 2.4). The lowest part is built with aggregate of broken brick. Then the aggregate changes to alternate layers of brick and tufa (a porous volcanic rock). The upper part of the dome is built with alternate layers of tufa and pumice, the latter especially imported from Mount Vesuvius to reduce the weight of concrete.

The inside of the Pantheon shows the marks of the wooden formwork.

Figure 2.1   The Pantheon in Rome, built approximately A.D. 123 and still in use. The span of 143 ft (44 m) was not reached again until the fifteenth century (138 ft or 42 m in the Florence Duomo) and was not surpassed in a building prior to the nineteenth century.   The construction is massive; the walls are 23 ft (7 m) thick between relieving arches.

However, the outside, like most Roman concrete structures, is faced with brick, placed both as permanent formwork and as a veneer. The Romans rarely used exposed concrete on the outside of buildings.

2.2 REDISCOVERY OF CONCRETE IN THE EIGHTEENTH CENTURY

The use of concrete as a major load-bearing material declined rapidly after the third century A.D., but the technique survived to some extent in both the Eastern and the Western Empires, where a core of rubble laid in lime mortar continued to be used with a stone or brick facing. However, the outer facing became more and more the load-bearing element and the rubble core a mere filling. With the refinement of wall thickness during the Gothic era, the rubble core assumed an increasingly subordinate role. In the German Rhineland, which had deposits of pozzolana (called Rhenish trass), the use of hydraulic mortar continued right through the Middle Ages; however, the masonry blocks were laid with mortar joints (in the manner still used today), whereas in the Roman *opus caementitium* the mortar was cast.

There is thus a gap of more than a thousand years between the great era of Roman concrete and its modern rediscovery, which is generally credited to John Smeaton (Refs. 2.6 and 2.7).

Smeaton's work was commissioned by the Brethren of Trinity House, a fraternity of seamen founded during the reign of Henry VIII, which had been transformed into a body of lighthouse commissioners. During the eighteenth century several attempts were made to build a lighthouse on the Eddystone, an exposed rock off the shore of Plymouth in England, on the principal shipping route to North America. Because the rock was frequently awash, lime mortar was washed out from the masonry joints, and because of the rough weather two timber lighthouses were destroyed in quick succession. Smeaton correctly concluded that the pure white lime commonly used for mortar was inferior in its hydraulic properties (i.e., its ability to harden under water) to gray lime, which contained some clay impurities. He further observed that pozzolana (*pulvis puteolanus* and Rhenish trass) derived its hydraulic properties from the combination of lime (calcium oxide) and clay (aluminum silicate). In the end he lacked the courage of his convictions and, in addition to using hydraulic mortar in the new Eddystone Lighthouse, he also cut the stones with mortise and tenon joints to ensure the cohesion of the masonry.

However, the Smeaton report of 1791 aroused others to seek a revivial of Roman concrete. James Parker in 1796 patented "Roman Cement," a natural cement (i.e., a natural mixture of lime and aluminum silicate) ground from chalk nodules in the London clay, and a variety of other patents followed.

In 1824, Joseph Aspdin patented "Portland Cement," an artificial cement made from a mixture of limestone and clay obtained from nearby, but different sites, and this material has been used with only minor changes since. The important innovation is the use of limestone or chalk, and clay or shale from different quarries; since both materials are plentiful, this freed cement manufacture from the need to find both components in a single deposit. At the time, Portland stone (a fine-grained limestone from the Isle of Portland) was the most favored building stone in London, and it was noted for its durability. Artificial cement made from limestone and clay is still called portland cement in every language. The first portland cement factory in the United States opened in 1871.

2.3 REINFORCED CONCRETE

The concept of reinforcing masonry and concrete also has a prehistory. Clamps of bronze or iron are found from time to time in various parts of the world. The Romans used them frequently when they encountered excessive tension, particularly in arches. If the stone blocks on the tension face are tied together with U-shaped pieces of iron, the joints cannot open up, and the load-bearing capacity is thus increased.

In the Renaissance the technique was extended to build domes far thinner than the Roman concrete domes; Brunelleschi's famous dome in Florence has several "chains" consisting of blocks of hard sandstone joined with iron clamps (Ref. 2.8).

The idea of increasing the bending strength of concrete by adding reinforcement occurred simultaneously to a number of people in Europe, who took out patents as follows: William B. Wilkinson (Fig. 2.2) of Newcastle-upon-Tyne, England (1854); and in France, Joseph-Louis Lambot (1855), François Coignet (1861), and Joseph Monier (1867).

Although telegraphic communication between the United States and Europe had been established by submarine cable in 1866, after several unsuccessful attempts, there was still a communication gap and the American development of reinforced concrete was essentially separate from that of Europe.

An early example was a house that William E. Ward built for himself in 1873 at Port Chester, a suburb of New York. Ward was a mechanical engineer who owned a factory for making bolts. Fire was a common occurrence in the timber-floored (and frequently all timber) houses of the New York region, which has hot summers and cold winters that make heating essential. Ward was determined to build a fireproof residence. After he saw a copy of *A practical treatise on Coignet-Beton and other artificial stone*, published in New York in 1871 by Q. A. Gillmore, he performed a series of experiments on concrete beams reinforced with iron bars. He first ascer-

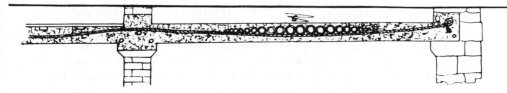

Figure 2.2 W. B. Wilkinson's patented reinforced concrete floor. The main reinforcement consisted of discarded colliery cables fixed at the top of the beams, and draped toward midspan so that the shape of the reinforcement conformed to that appropriate to a continuous beam. Drain pipes were inserted near midspan to reduce the weight. Although Wilkinson lived to the age of 83, and his successful building firm survived him, his influence on reinforced concrete design was slight. Indeed, his patent had been forgotten when, in 1955, the contractor demolishing a building erected by Wilkinson in 1865 noticed the unusual nature of the construction and called in Professor Fisher Cassie of Newcastle University, who subsequently located several more Wilkinson buildings (Ref. 2.16).

tained that the reinforcement should be at the bottom and then conducted tests on shear resistance, deflection, and fire resistance.

In 1872, he commissioned Robert Mook, a New York architect, to design a house that would contain virtually no timber. It was built in a style modeled on the French Renaissance, and common for villas in the Hudson Valley, with a mansard roof, two crenelated towers, and an open veranda with Tuscan Columns. The only significant amount of wood was in the window frames, doors, and rails for the stairs. The mansard roof was concrete reinforced with iron rods, and the carpets were fixed to nailing strips cast in the concrete floors.

Ward did his own structural design and personally supervised the work. The floors were mostly $3\frac{1}{2}$ in. (90 mm) thick, reinforced with rods $\frac{5}{16}$ in. (8 mm) in diameter and with spans as long as 6 ft (1.8 m). The hollow veranda columns acted as downpipes for the rainwater. All the moldings and structural coffered ceilings were cast in place with concrete made of portland cement, sand, and a crushed bluestone. The house is still standing. In 1877, *The American Architect* described it and published several articles on the use of concrete. Ward himself wrote a paper *Beton in combination with iron as a building material*, which appeared in Volume 4 (1882–1883) of the *Transactions of the American Society of Mechanical Engineers*.

Of more significance was the experimental work undertaken by Thaddeus Hyatt, starting in 1855. In 1877 he published for private circulation *An account of some experiments with Portland-Cement-Concrete combined with iron as a building material, with reference to economy of metal in construction, and for security against fire in the making of floors, roofs and walking surfaces*. This paper contained data from tests of about 50 beams for strength and fire resistance and a method of calculation.   Hyatt took out several patents in the United States and England.

Ernest Leslie Ransome was born in Ipswich, England, in 1844.   In the same year his father started a factory for making precast concrete blocks.   In 1866, the son was sent to a branch established by the senior Ransome in Baltimore, but in 1870 he was already working independently.   It is not known how much Ransome had heard of Ward's and Hyatt's experiments or any European work; in 1884, however, he patented a spiral-twisted square bar to improve the bond between iron and concrete.   From 1889 to 1891 he built the Leland Stanford Junior Museum near San Francisco at the newly founded Stanford University, a classic three-story fireproof building in which no timber was used.   The floors were reinforced concrete, the concrete tiles on the roof were supported on interlocking iron trusses, the windows were metal frames, and the hall was marble-surfaced.   Ransome was probably the first to remove the cement film on the surface of concrete by tooling to expose the aggregate. For a nineteenth-century concrete building the finish was exceptionally good.

Ransome developed his idea on reinforced concrete frame construction on several multistory buildings before erecting in 1902 the first reinforced concrete skyscraper, the 16-story Ingalls Building in Cincinnati, which was 177 ft (54 m) high.   This building was cast with a frame in which the walls acted merely as curtain walls.   It was a functional building in the tradition of the Chicago School.   In the same year Ransome patented his system of concrete-iron frame construction, for which he claimed, as the main advantage, that the frame structure obviated the need for thick concrete walls, a method that made large windows possible wherever required.

2.4 THEORY OF REINFORCED CONCRETE

Most of the early work on the theory of reinforced concrete design took place in France and Germany.   There were three important problems to be solved. One was the behavior of the concrete structure as a rigid frame.   Ransome established the principle at the turn of the century in America, and François Hennebique did so at the same time in Europe.

Hennebique, born in 1842, became a master carpenter and worked for many years on the restoration of medieval cathedrals.   He first used concrete reinforced with iron bars to construct fireproof floors in 1879 and conceived the idea of bending up the reinforcement to resist the tension developed in the concrete over the supports.   He patented this process in 1892 (Fig. 2.3). Hennebique's building firm gradually did more and more work in reinforced concrete.   In 1892, he closed down his construction firm and set up a consulting engineering office; the construction was done by contractors licensed and trained by the Hennebique organization.

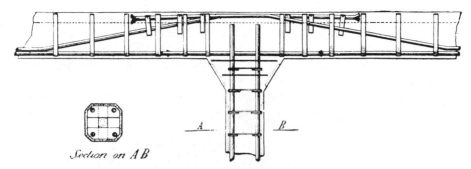

Figure 2.3 Arrangement of the reinforcement in the Hennebique system, patented in 1892, and registered in the United Kingdom and its colonies in 1897 (Patent No. 30143). The opening clause of the specification states the advantages claimed: "The use of strengthened beton in buildings has within recent years greatly developed. It has been thought possible by mixing beton and iron or steel to replace the purely metallic elements of building construction by parts equally incombustible but lighter and more simply and rapidly made. In any case the mixture of cement or hydraulic lime which resists perfectly compression, with iron or steel which most particularly resists tension and flexion, has not hitherto been capable of being carried out in a judicious and rational manner."

Hennebique emphasized that reinforced concrete was not a cheap substitute material. He stressed the need for constant supervision to achieve good workmanship and was the first to substitute steel for wrought iron as reinforcement.

Like Ransome, he regarded the concrete structure as a frame consisting of columns and floors; the external walls carried no load and did not need to be thick. Lightness was a quality evident in all of Hennibique's structures and may be one reason why they compared favorably in cost with others built at the turn of the century.

Among Hennebique's designs was a curved freestanding staircase for the Petit Palais de Champs-Élysées, built in 1898 for the Paris Exhibition of 1900 (the apparently supporting walls were built subsequently to provide storage space). The design was largely intuitive; 20 years later when the torsional strength of reinforced concrete was better understoood, few engineers would have dared to build this light structure.

Louis Gustave Mouchel, Hennebique's representative in England, designed the 11-story skeleton frame of the Royal Liver Building in Liverpool and set a European record for a tall building in reinforced concrete. Built in 1908 and 1909, its height is 167 ft (51 m) to the main roof, but this is topped by a clock tower and a giant effigy of the legendary liver bird, which takes the total height to 310 ft (95 m). It was the first British skyscraper.

Hennebique introduced the rigid-frame concept for reinforced concrete structures in Europe but did not design his floor structures as continuous beams and slabs, although the theory had been published 40 years earlier. Instead, he used coefficients for the bending moment at the supports and midspan ranging from $\frac{1}{8}$ to $\frac{1}{30}$; that is, the bending moment was taken as $W\ell/8$, and so on, where W is the total load carried by the beam and ℓ is its span. These coefficients appear in most early textbooks on reinforced concrete (Ref. 2.9) and in early building regulations for concrete construction (Ref. 2.19). Reinforced concrete floors have been designed as continuous structures only since 1910.

A method proposed by Mattias Koenen in a book *Das System Monier*, published by Wayss and Freytag in 1886, became the basis of reinforced concrete design. Koenen proposed that Hooke's law be assumed to apply to concrete as well as iron, that the iron be assumed to resist the whole of the tension, and that perfect adhesion be assumed between the iron and the concrete. He also noted that the coefficient of thermal expansion of iron and concrete was nearly the same and consequently no stresses due to changes of temperature would be set up between them; this was essential to a fireproof material. Koenen, however, placed the neutral axis at half depth, which ignored the difference in the moduli of elasticity of iron and concrete.

This mistake was corrected in a paper presented to the *Société des Ingénieurs Civils de France* in 1894 by Edmond Coignet and Napoleon de Tedesco. In 1887, Charles Rabut gave the first course of reinforced concrete design at the *École des Ponts et Chaussées* in Paris, and in 1899, in Liége, Paul Cristophe published *Le Béton armé et ses applications*, which was based on the straight-line theory of Coignet and Tedesco. Charles F. Marsh, in the first English-language book on reinforced concrete design (Ref. 2.9), in turn used Cristophe's method.

The theory of reinforced concrete design was greatly simplified by Emil Mörsch, chief engineer at Wayss and Freytag, in a book *Der Betoneisenbau* (later renamed *Der Eisenbetonbau*, Ref. 2.22), published by the company in 1902. (Mörsch became professor at the Federal Institute of Technology in Zurich in 1904.) This remained the standard method until the 1970s, when ultimate strength design was introduced in most countries.

At the turn of the century Hennebique had become "le Napoléon du Béton Armé," but in 1901 he suffered a serious setback when the five-story structure of the *Hotel zum Goldenen Bären* in Basel collapsed during construction with loss of life (Ref. 2.5, p. 124; Ref. 5.16, p. 62). The commission of inquiry, under the chairmanship of Professor W. Ritter of the Zurich Technical University, blamed faults in the design by the Hennebique organization in Paris and poor workmanship by the local contractor. In particular, it criticized the use of unwashed sand and gravel, including some taken directly from the building site, and the failure to test the quality of the cement and the compressive strength of the concrete.

The collapse led in 1903 to the introduction in Switzerland of the first building regulations for reinforced concrete, and other countries soon fol-

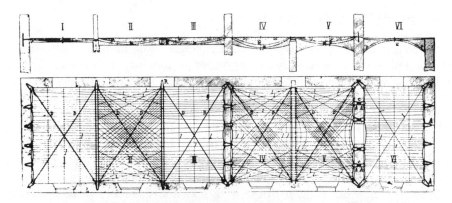

Figure 2.4 The Matrai system, originally patented in Hungary. Professor Matrai likened his system to a spider's web firmly attached to the points of support. The system required "three or four times less metal" than ordinary reinforced concrete beams. "In bays marked I and III the wires are placed so as to distribute the load equally over the beams, which only support half the load, being assisted by diagonal cables which are calculated to take the other half. In bays II, IV, and V the diagonal cables are not sufficient to transmit half the load to the extremities of the beams, and they are supplemented by diagonal wires. The transverse wires in bay VI are sufficient to distribute the load equally over the walls, and the diagonals and longitudinals are only employed to intertie the transverse wires." The upper surface of the concrete followed the curve of the wires, and the hollow was filled with coke breeze concrete.

lowed.   In the United States a Joint Committee set up by the American Society of Civil Engineers produced its first code in 1908.

The collapse also highlighted the third important problem in reinforced concrete design: the need for quality control of a material as variable as concrete (see Section 5.6). In the first two decades of this century concrete-mix proportioning was put on a scientific basis. In 1918, Professor Duff Abrams of the Lewis Institute in Chicago enunciated the rule that the strength of concrete is inversely proportional to the water/cement ratio, that is, the ratio that the amount of water bears to the amount of cement in the mix. The ancient Romans knew that good concrete needed as little water as possible, but this rule supplied a quantitative basis.

Until 1919, reinforced concrete was largely dominated by the various patent systems. Marsh (Ref. 2.9) in 1904 listed 43 different systems, of which 15 were patented in France, 14 in Germany or Austria-Hungary, 8 in the United States, 3 in Britain, and 3 elsewhere. These systems varied from the exceedingly complex (Fig. 2.4) to simple arrangements similar to those employed today (Fig. 2.3)

The latter triumphed because even in the early twentieth century the extra cost of labor (including extra supervision because of the complexity) did not compensate for the material saved.

2.5 ARCHITECTURAL CONCRETE

The triumphant progress of reinforced concrete from its hesitant beginning in the mid-nineteenth century to its present prominence in long-span and, to a lesser extent, multistory buildings was held back by two problems.   One was the lack of satisfactory design theory for statically indeterminate structures and the other was the need for a better method of finishing concrete surfaces.

New concrete can be made to look similar to stucco, highly esteemed in the nineteenth century, particularly if it is troweled to a smooth surface. The gray color can be masked by paint.   However, the smooth finish produced by lengthy troweling is due to an excess of cement being worked up to the surface.   When the cement dries, it shrinks and produces surface crazing, and sometimes large cracks, if the contraction of the concrete is restrained. Paint can cover surface crazing but not deep cracks.   Moreover, the paints available in the early years of this century were not durable on concrete and required frequent renewal.

The late nineteenth and the early twentieth centuries were notable for eclectic revivals, some of which were unsuitable for concrete.   Concrete structures in the style of the early Renaissance or Greek Revival performed reasonably satisfactorily, but buildings with sculptural decorations, such as the Neo-Gothic, were troubled by cracking due to the stress concentrations that inevitably occur at the reentrant corners.   Casting 20 concrete gargoyles from a single mold instead of sculpting them individually from natural stone proved a great economy, but unlike natural stone the concrete did not age gracefully.

Although concrete was a material much admired by some of the pioneers of the modern school, their use of it often showed a lack of appreciation of its limitations.   Le Corbusier in *Towards a New Architecture* illustrated some of his concrete houses and underneath wrote this caption:

> The concrete was poured from above as you would fill a bottle.   A house can be completed in three days.   It comes out from the shuttering like a casting. But this shocks our contemporary architects, who cannot believe in a house that is made in three days; we must take a year to build it, and we must have pointed roofs, dormers, and mansards.   (Ref. 3.5, p. 212)

Le Corbusier's client, Henry Frugès, described the construction of the concrete house at Pessac in 1926–1927:

> A further difference of opinion between Le Corbusier and myself: with his inveterate hatred of all forms of decoration (which stemmed from his Protestant background and the general austerity of his personality) he wanted to leave the walls completely unfinished so that they still showed the marks of the shuttering. I was flabbergasted.   He told me that if we wished to offer the houses at the

lowest possible price, we could not afford to spend money on unnecessary
luxuries. He then launched into a diatribe against ornamentation exclaiming
"We are tired of décor, what we need is a good visual laxative! Bare walls,
total simplicity, that is how we restore our visual sense"! I understood him
only too well, because we both wanted to build economically, but he did not
understand me. (Ref. 3.7, p. 9)

The garden suburb of Pessac was not a success. It was an interesting experiment because it provided an early exercise in standardization, but the
concrete weathered badly. Similar problems were encountered in buildings
designed by other modern architects in the 1920s and 1930s.

Auguste Perret, who was 13 years older than Le Corbusier, was the
first architect to produce a high-quality finish on exposed concrete. He
achieved it by acting both as designer and builder and by paying careful
attention to detail. Perret's formwork carpentry was always immaculate and
his concreting operation was properly controlled. His buildings are still in
excellent condition (Ref. 3.4). Perret made extensive use of precast concrete
blocks and used them for pierced screen walls in conjunction with glass. The
earliest large-scale use of this technique was in the Church of Notre Dame
at Le Raincy, a suburb of Paris, constructed in 1922; its walls consisted largely
of glass set in concrete.

Perret also appreciated the ease with which arches, vaults, and staircases
could be built in reinforced concrete. Many of his buildings have elegantly
curved stairs. In the Esders Clothing Factory, built in Paris in 1919, Perret
used semicircular arches to support the roof at midspan, thus doubling the
clear span of the beams carrying the roof without a central line of columns.
Apart from the functional advantage of the long span, the arches created
visual interest in an otherwise plain interior.

Most of Perret's reinforced concrete frames emphasized the horizontal
and the vertical members, and the columns were usually shaped in accordance
with Vitruvius' rules of entasis. Perret was a great innovator but not a
revolutionary like Le Corbusier, and he retained his loyalty to classicism even
in his late buildings in the 1930s and 1940s. The great increase in the cost
of labor in the years following World War II made Perret's methods of construction obsolete. It was no longer economically feasible to build precision
formwork for site-cast concrete and then to fit precisely precast concrete
blocks between the site-cast columns.

Among the great architects of the early twentieth century the one most
percipient of the qualities of concrete was Frank Lloyd Wright, although the
material was never so important in his work as in Perret's or Le Corbusier's.
Wright used concrete for the first time in 1905 in the E-Z factory in Chicago,
but the reinforced concrete frame was faced with brick. In the Unitarian
Church in Oak Park, Illinois, built in 1906, the entire facade was made of
site-cast concrete with an exposed pebble-aggregate finish. In the 1920s

Wright used precast concrete blocks in a number of buildings. In some he used them as permanent formwork for the reinforced concrete frame, thus solving the problem that had defeated Le Corbusier: how to produce an acceptable concrete finish on vertical surfaces. At the same time he avoided the high labor cost of Perret's precision-made formwork.

In the 1930s concrete was still an engineer's rather than an architect's material. Even cement manufacturers in their publicity tended to stress "the ease with which concrete can be veneered with natural stone, producing a building which is indistinguishable from one built from natural stone, but so much cheaper." The problem of the surface finish of concrete was solved only in the 1950s and 1960s (see Chapter 3).

EXERCISES

1. Read something about Roman concrete. The three books by Elizabeth Blake are the most detailed treatise on the subject (Ref. 2.4), but any book on the history of Roman architecture or Roman engineering has some sections and illustrations of Roman concrete construction.

2. If you have access to a large library or an old, established library, you may be able to find an early book (i.e., one published before 1918) on reinforced concrete design or reinforced concrete construction. [The first book in the English language was published in 1904 (Ref. 2.9).] Compare the illustrations with those shown in this book (or any other modern book on reinforced concrete) and note the differences.

3. Look at the Anniversary Issue of the *Journal of the American Concrete Institute* (Ref. 2.19). It contains an excellent history of reinforced concrete in America. Alternatively, look at an account of the history of reinforced concrete in Europe, such as Refs. 2.17 or 2.18.

REFERENCES

2.1. M. Goldfinger, *Islands in the Sun*, Lund Humphries, London, 1969, pp. 40–49.
2.2. Vitruvius, *The Ten Books of Architecture*, trans. M. H. Morgan, Dover, New York, 1960, Book 2, Chap. 6, pp. 46–47.
2.3. H. Plommer, *Vitruvius and Later Roman Building Manuals*, Cambridge University Press, London, 1973.
2.4. Marion Elizabeth Blake, *Ancient Roman Construction in Italy from the Prehistoric Period to Augustus*, Carnegie Institution of Washington Publication 570, Washington, D.C., 1947; *Roman Construction from Tiberius through the Flavians*, Carnegie Institution, Publication 616, Washington, D.C., 1959; *Roman Construction in Italy from Nerva through the Antonines*, American Philosophical Society, Memoirs, Vol. 96, Philadelphia, 1973.
2.5. Gustave Haegermann, Günter Huberti, and Hans Möll, *Vom Caementum zum Spannbeton*, Vol. I., Bauverlag, Wiesbaden, 1964.
2.6. Henry J. Cowan, *An Historical Outline of Architectural Science*, 2nd ed., American Elsevier, New York, 1977, pp. 46–55.

2.7. Samuel Smiles, *Lives of the Engineers—Smeaton and Rennie*, John Murray, London, 1904, pp. 1–123.

2.8. F. D. Prager and G. Scaglia, *Brunelleschi*, MIT Press, Cambridge, Mass., 1970, pp. 33–38.

2.9. Charles F. Marsh, *Reinforced Concrete*, Constable and Co., London, 1904.

2.10. D. A. L. Saunders, "The Reinforced Concrete Dome of the Melbourne Public Library, 1911," *Architectural Science Review*, Vol. 2, 1959, pp. 34–46.

2.11. "Report of the ASCE–ACI Joint Committee on Ultimate Strength Design," *Proc. American Society for Civil Engineers*, Vol. 81, 1955, Paper No. 809.

2.12. H. J. Cowan, "Ancient Roman Concrete," *Journal of the Royal Institute of British Architects*, Vol. 61, 1954, p. 120.

2.13. A. C. Davis, *A Hundred Years of Portland Cement*, Concrete Publications, London, 1924.

2.14. P. Gooding and P. E. Halstead, "The Early Years of Cement in England," Paper No. 1, *Third International Symposium on Chemistry of Cement*, London, 1952.

2.15. J. O. Draffin, *A Brief History of Lime, Cement, Concrete, and Reinforced Concrete*, University of Illinois Engineering Experiment Station, Reprint No. 27, 1943.

2.16. W. F. Cassie, "Early Reinforced Concrete in Newcastle-upon-Tyne," *Structural Engineer*, Vol. 33, 1955, pp. 134–137.

2.17. S. B. Hamilton, *A Note on the History of Reinforced Concrete in Buildings*, National Building Studies, Special Report No. 24, Her Majesty's Stationery Office, London, 1956.

2.18. Fiftieth Anniversary Number, *Concrete and Constructional Engineering*, Vol. 51, 1956, pp. 1–262.

2.19. Anniversary Issue, *Journal of the American Concrete Institute*, Vol. 25, 1954, pp. 409–524.

2.20. D. E. Bornemann, "Concrete and Reinforced Concrete Construction since the Formation of the German Committee for Reinforced Concrete" (in German), *Beton und Stahlbetonbau*, Vol. 53, 1958, pp. 29–33.

2.21. F. Emperger (ed.), "The Elements of the Historical Development of Reinforced Concrete, Experiments and Theory" (in German), *Handbuch für Stahlbetonbau*, 3rd ed., Vol. I, W. Ernst, Berlin, 1921.

2.22. Emil Mörsch, *Concrete Steel Construction* (English transl. of *Der Eisenbeton-bau*), Engineering News Publishing Co., New York, 1909.

Chapter 3

The Use of Concrete in Architecture

This chapter examines briefly the architectural problems in the use of concrete. We first consider the various methods of treating the surface of concrete, their advantages and their limitations. In this respect there have been important innovations in the last few years. We then examine the use of concrete as a plastic, three-dimensional material.

3.1 SURFACE FINISH OF CONCRETE

Concrete troweled to a smooth surface has, after hardening, a dull-gray appearance. This does not improve with weathering, because fine cracks, formed by shrinkage, collect dirt. By contrast, most natural stones age gracefully.

It is sometimes said that the contrast between concrete and stone is between an artificial and a natural material. However, many types of brick (an artificial material) age well, whereas mud, a natural material widely used for building in Africa and Asia, requires special surface treatment for durability.

The problems with concrete surfaces arise from the gray color and the chemical and physical changes in portland cement. Surface treatment can therefore take four forms:

1. Covering the cement
2. Changing the color of the cement

3. Removing the cement

4. Masking the disfigurement of the cement surface by a stronger pattern

3.2 PAINTING

Cement can be covered with a suitable paint, which is either cement-based or has an alkali-resistant vehicle such as polyvinyl acetate or acrylic. On a long-term basis this is quite expensive, because the paint needs renewing every few years. It is noteworthy that some early painted concrete buildings have turned into slums partly because of inadequate attention to repainting (see Section 3.8).

Figure 3.1 Multistory building with a site-cast reinforced concrete frame and precast cladding panels. The panels are cast in steel forms using white cement, and their surface is acid-etched after hardening. (Central Square in Sydney, completed in 1970; architects: Fox and Associates; structural engineer: Civil and Civic Pty Ltd; photograph: Max Dupain.)

3.3 WHITE AND COLORED CEMENT

Changing the uninteresting gray color of portland cement is not in itself
sufficient, and the use of colored cement must generally be supplemented by
exposed aggregate or some masking treatment (see Sections 3.4 to 3.6).   The
most important change of color is from gray to white (Figs. 3.1 and 3.11).

Pure white cement is produced from china clay (instead of colored clay)
and pure white limestone, fired with oil (instead of powdered coal), and
ground with pebbles or nickel alloy balls (instead of iron balls) (Ref. 3.1).
It is therefore appreciably more expensive than ordinary gray cement.   Off-
white cement is a cheaper intermediate product.

Colored cement can be produced by adding mineral pigments; the lighter
shades require off-white cement.   There is a good range of red, yellow,
brown, and green pigments suitable for mixing with cement.   However, color
in concrete has been more commonly produced by using colored aggregates.

3.4 EXPOSED AGGREGATE

Removal of the cement exposes the aggregate, which, if properly chosen,
should weather like good natural stone.

Figure 3.2   Site-cast concrete surface with 3/8-in. aggregate exposed by washing
with water.

Figure 3.3 Precast concrete panel cast "face up."   The aggregate is sprinkled on the surface after the concrete is cast, but while it was still wet, and rolled into the surface.

Aggregate can be exposed on site-cast concrete by washing the cement off the surface while the concrete is still green (Fig. 3.2).   The formwork is sometimes coated with a retarder to delay setting of the surface layer of the concrete.

Precast concrete wall panels are cast horizontally, either "face down" or "face up."   In face-down panels, a specially selected aggregate is placed at the bottom of the mold (the mold is sometimes coated with a retarder); the cement is then washed off or acid-etched to expose the aggregate.

In the face-up process the concrete panel is cast horizontally with ordinary aggregate, and the facing aggregate is sprinkled on top and rolled into the surface (Fig. 3.3).   Natural crushed aggregates are obtainable in crystalline white (quartz), dull white (limestone), pink, red, yellow, light and

TABLE 3.1 Maximum Distance at Which
Aggregate Texture Is Visible

Size of Aggregate Particles (in.)	Distance (ft)
$1\frac{1}{2}$	350
1	300
$\frac{3}{4}$	200
$\frac{1}{2}$	115
$\frac{3}{8}$	75
$\frac{1}{4}$	60

dark brown, light and dark green, and black. Broken glass can be used in almost any color where sharp edges are admissible. In addition, large or small pebbles can be used. Large pebbles give a rustic effect which acts as a good background for an interior or exterior garden. Broken-glass aggregate combines well with display lighting in buildings devoted to shopping or entertainment.

Exposed aggregate is not unduly expensive, and it is a good material for interiors and for wall surfaces of limited height (Ref. 3.2).

For exteriors that may be seen both close up and from a distance, readability of the aggregate becomes a problem. J. G. Wilson (Ref. 3.3) of the (British) Cement and Concrete Association has estimated the maximum distance at which the texture of exposed aggregate can be discerned (Table 3.1).

Three-eighth-inch chips of varied color, commonly used for exposed aggregate finishes, are evidently not suitable for buildings taller than 75 feet. An aggregate with a mixture of black and white (Fig. 3.4) or colored aggregate, however delightful at close quarters, looks an indifferent gray from a distance. The problem cannot be solved by using large pebbles (2 to 4 in.) because these would produce a finish unacceptable at close quarters.

A possible solution is the use of white or off-white cement with white aggregate (Fig. 3.1), but the recent trend toward sculptured finishes is largely due to the "readability" problem posed by exposed aggregates at a distance.

3.5 BROKEN SURFACES

Breaking part of the concrete surface locally exposes the aggregate and displays a rough texture that distracts attention from minor blemishes, particularly if the sun throws irregular shadows (Fig. 3.5). This can be done by casting the concrete against ribbed timber forms, and breaking the surface off the ribs with blows from an ordinary hammer (Fig. 3.5), or by casting

Figure 3.4 Site-cast concrete structure with permanent formwork of precast concrete. The precast units have a surface of exposed aggregate, a mixture of 3/8-in. black-and-white particles. This looks excellent at close quarters, but appears gray from a greater distance because it is difficult to distinguish the two colors at a distance of more than 75 ft. (see Table 3.1). (Australia Square Tower, Sydney; design and construction: Civil and Civic Pty. Ltd.; architect: Harry Seidler; photograph: Max Dupain.)

ropes into the concrete near the surface and pulling them out to leave rope-textured grooves between broken surfaces (Fig. 3.6); both techniques are applicable to site-cast concrete. An alternative approach is to split precast panels and use the interior break as the exposed surface.

These methods are economical, and produce finishes which from a distance look better than most exposed aggregate surfaces. Nearby they may

Figure 3.5 Hammered and ribbed finish. The concrete is cast against ribbed timber forms, and the tops of the ribs are subsequently broken off with blows from a hammer.

Figure 3.6 Patterns produced by ropes cast into concrete. These are cast near the surface into site-cast concrete and pulled out, leaving textured grooves between broken surfaces.

look brutal, but few people see them that way if the entrance and interiors are appropriately treated to be seen at close quarters.

3.6 BOARD-MARKED SURFACES

Board-marked finishes are effective because the pattern left on the concrete by the timber forms is sufficient to distract attention from the minor blemishes inevitable on concrete surfaces.

It is frequently assumed that this is cheap because the formwork is required in any case; but due to the high labor content of carefully made formwork, this is one of the more expensive finishes (Fig. 3.7).

This surface treatment was used by Auguste Perret (see Fig. 3.9) in the earlier years of this century when carpenters' wagers were much lower. Le Corbusier often employed roughly made formwork, whose markings would have been criticized if used by a man of less distinction.

Figure 3.7   Board-marked surfaces formed by rough-sawn timber, carefully weathered to emphasize the growth rings. The structure consists of reinforced concrete slabs supported on post-tensioned beams (Chapter 15); the projections are metal covers over the prestressing anchorages.

Figure 3.8 Symbolic pattern on site-cast concrete formed with timber molds. (Maccabean Hall, Darlinghurst, N.S.W.; architect: Henry Epstein; sculptor: Lyndon Dadswell; photograph: Kerry Dundas.)

3.7 SCULPTURED SURFACES

Stronger geometric, decorative, or symbolic patterns (Fig. 3.8) can be applied either to precast panels or to site-cast concrete with form liners of plastics or timber. Evidently, they can be used only on a restricted scale, or for buildings of special significance, partly because of the cost and partly because of the overpowering effect of too much sculptured concrete.

3.8 CONCRETE AS A PLASTIC THREE-DIMENSIONAL MATERIAL

When a new material comes into use, there is a natural tendency to employ techniques developed for older materials. The early use of plain concrete in architecture is contemporary with highly decorative styles such as Neo-

Gothic.   At the time it was tempting to produce concrete castings from a mold to save the expense of stone carving, but the result was not acceptable even in a church for the working class, partly because of the cement color and partly because of the cracks formed at the reentrant corners, which collected dirt and in some instances led to complete disintegration.

For the same reason imitations of cast iron, then at the peak of its popularity, were not successful.   Both materials are cast into a mold, but the sharp and intricate detail of cast iron was not reproducible in concrete.

Concrete might have fared better if its initial contact with architecture had been earlier, during the Greek Revival, or later during the period of Art Nouveau, since that style could have been reproduced in concrete with less risk of failure.

By the turn of the century concrete had become discredited as an architectural material.   It was evidently an engineering material, admirable for lining sewers, good for structural frames because it was fireproof, but not to be shown on the facade of the building.

The credit for developing the architectural potential of concrete belongs mainly to Auguste Perret (Ref. 3.4) who appreciated the significance of strong, simple outlines rather than intricate detail, a surface texture formed by board marks, and the potential for the interaction between daylight and repetitive concrete castings (Fig. 3.9), which was later adapted successfully to ventilating sun screens in the tropics (Ref. 3.14).

The writings of Le Corbusier (Ref. 3.5) and other theorists (Ref. 3.6) have also had much influence on present-day practice, but the actual performance of concrete in early buildings by modern architects, such as Le Corbusier, fell far short of expectations (Ref. 3.7) because the plain, smoothly

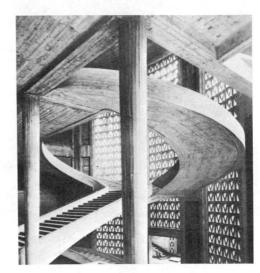

Figure 3.9 Staircase in the Museum of Public Works, Paris, 1939, a late work by Auguste Perret (architect and builder).

Figure 3.10 Coal-loading dock at a power station in Civitavécchia, Italy, completed in 1955: an easily readable structural feature in an otherwise unattractive setting. (Architect: Riccardo Morandi, Ref. 3.11.)

finished concrete surfaces used in the 1920s weathered badly. In some instances concrete surfaces were painted white, and the paint (which was not always of a suitable type) was not renewed as often as necessary.

Some of the best modern designs have relied on an architectural expression of the structure (Figs. 3.9 to 3.16, and Refs. 3.8 to 3.13). However, it does not follow that a correctly designed structure is automatically beautiful.

Figure 3.11 Audience Hall at the Vatican, Rome, completed in 1972. The splayed columns, site-cast with white marble chips and white cement and subsequently bush-hammered, support the curved roof structure. (Architect and builder: Pier Luigi Nervi.)

Figure 3.12 Prestressed concrete footbridge (see Chapter 15) inside the East Building of the National Gallery of Art in Washington, a spectacular structural feature spanning 135 ft (41 m) with a depth of only 4 ft (1.2 m). (Architect: I. M. Pei and Partners; structural engineer: Weiskopf and Pickworth; photograph courtesy of Weiskopf and Pickworth.)

Figure 3.13 Gladesville Bridge, Sydney, opened in 1964, the world's longest-spanning concrete arch (1000 ft or 300 m). The bridge is built from concrete segments with only nominal reinforcement, and lifted off the formwork with prestressing jacks, so that the concrete is prestressed against the bridge's abutments (see Fig. 15.1). (Designer: Maunsell and Partners Pty. Ltd.; photograph courtesy of the N.S.W. Department of Main Roads.)

Figure 3.14   CN Tower in Toronto, Canada, opened in 1976, the world's tallest tower.   It has a height of 1815 ft (533 m); for comparison, the World Trade Center in New York has a height of 1350 ft (411 m) and the Eiffel Tower in Paris 984 ft (300 m).   It contains broadcasting and telecommunication facilities, restaurants, and observation decks.   The structure is of post-tensioned concrete.   (Photograph courtesy of Canadian National.)

Figure 3.15   Assembly Hall at the University of Illinois, Urbana.   Its roof has a folded-plate dome with a post-tensioned compression ring, completed in 1963.   The dome has a diameter of 400 ft (122 m) and an average thickness of concrete of 7 in. (175 mm).   (Architect: Harrison and Abramovitz; structural engineer: Amman and Whitney; photograph courtesy of Harrison and Abramovitz.)

Figure 3.16   King County Stadium, Seattle, Washington, completed in 1975.   Its roof is formed by concrete ribs and thin shell sections that have a hyperbolic-paraboloid-derived double curvature.   The dome has a span of 661 ft (202 m), and the thin-shell segments have an average thickness of 5 in. (125 mm).   (Architects and structural engineers: Naramore, Skilling, Prager; photograph courtesy of Skilling, Helle, Christiansen, Robertson.)

3.9 THE PROBLEM OF SPAN

Creating the longest span has exercised a peculiar fascination on clients and their architects throughout the ages.   In the eighteenth century, the age of the Grand Tours, gentlemen, including architectural gentlemen, would travel long distances in great discomfort to see a building that spanned farther than any they had so far inspected.

Concrete cannot compete with steel in long-span bridges.   The world's longest span in concrete is the Gladesville Bridge in Sydney (Fig. 3.13), which spands 1000 ft (300 m), as compared with 4260 ft (1 300 m) for the longest span in steel (the Verrazano Narrows Bridge in New York).

Concrete compares favorably with steel in long-span buildings, because the concrete provides the surface of the roof as well as the supporting structure.

Ancient Rome established a long-span record with the Pantheon (Fig. 2.1) which held for more than 1700 years (Table 3.2).   Brunelleschi's masonry dome for S. Maria del Fiore in Florence, and Michaelangelo's S. Pietro in Rome fell a little short of the span of the Pantheon, and no other building surpassed it until iron came into use in the nineteenth century.

The span of the Pantheon was exceeded by a modern concrete dome for the first time in 1912 (by the Centenary Hall in Wroclaw, then Breslau), and really long spans were attained for the first time in 1958 by Nicolas Esquillan's Palais de Centre National des Industries et des Techniques at Paris-La Défense (Ref. 3.12), which spans 677 ft (206 m).   Since then several concrete roofs of comparable size (Figs. 3.15 and 3.16) have been built in America.

TABLE 3.2   Notable Long-Span Buildings

Year of Completion (A.D.)	Name, Place, and Type of Roof	Span (ft)	Average Thickness of Shell or Combined Thickness of Double Shell (in.)	Ratio of Span to Thickness
123	Pantheon, Rome; solid concrete dome with relieving arches (Fig. 2.1)	143	156	11
1434	Duomo, Florence; double dome of masonry and brick	138	78	21
1710	St. Paul's, London; brick dome surmounted by brick cone: the lightest classical masonry dome	108	36	36
—	Large hen's egg	0.13	0.01	130
1924	Planetarium, Jena, East Germany, first thin reinforced concrete shell	82	$2\frac{1}{4}$	440
1953	Schwarzwaldhalle, Karlsruhe, West Germany, first prestressed concrete saddle shell	240	$2\frac{1}{4}$	1 300
1958	CNIT Exhibition Hall, Paris; double-reinforced concrete shell	677	$4\frac{3}{4}$	1 700
1975	King County Stadium, Seattle; hyperbolic paraboloid shell (Fig. 3.16)	661	5	1 600

EXERCISES

1. Look at a book on concrete surface finishes, for example, Ref. 3.2.

2. Look at a book on the architecture of Auguste Perret, the first architect to achieve a concrete surface finish of consistently high quality.   Collins (Ref. 3.4) gives a detailed and well-illustrated account of the work of Perret.

3. Look at some of the concrete buildings of Le Corbusier, which had great influence on the form and surface finishes of concrete structures, particularly in the 1950s and 1960s.   There is extensive literature on Le Corbusier, as well as his own books (such as Ref. 3.5).   Siegfried Giedion in *Space, Time and Architecture* (5th edition, Harvard University Press, Cambridge, Mass., 1967) was an uncritical admirer at times; Boudon (Ref. 3.7) has pointed out some of the major mistakes.

4. Read one or more books on the work of one of the great masters of the structural use of concrete, such as Maillart (Ref. 3.8), Nervi (Ref. 3.9), Toroja (Ref. 3.10), Morandi (Ref. 3.11), Esquillan (Ref. 3.12), or Candela (Ref. 3.13).

5. Look at the *Guinness Book of Records* on Structures, which lists the longest spans historically, by materials, and by structural use (John H. Stephens, *Structures*, Guinness Superlatives, Enfield, Middlesex, England, 1976.

REFERENCES

3.1. A. M. Neville, *Properties of Concrete*, Pitman, London, 1963.

3.2. H. L. Childe, *Concrete Finishes and Decoration*, Concrete Publications, London, 1964.

3.3. J. G. Wilson, *Concrete Facing Slabs*, Cement and Concrete Association, London, 1955.

3.4. Peter Collins, *Concrete, the Vision of a New Architecture*, Faber, London, 1959.

3.5. Le Corbusier, *Towards a New Architecture*, trans. F. Etchells, Architectural Press, London, 1970.

3.6. Ulrich Conrads, *Programmes and Manifestos on 20th Century Architecture*, Lund Humphreys, London, 1970.

3.7. Phillipe Boudon, *Lived-in Architecture — Le Corbusier's Pessac Revisited*, Lund Humphreys, London, 1972.

3.8. Max Bill, *Robert Maillart*, Pall Mall Press, London, 1969.

3.9. P. L. Nervi, *Structures*, trans. M. and G. Salvadori, Dodge, New York, 1956.

3.10. *The Structures of Eduardo Torroja — An Autobiography of Engineering Accomplishment*, Dodge, New York, 1958.

3.11. G. Boaga and B. Boni, *The Concrete Architecture of Riccardo Morandi*, Tirani, London, 1965.

3.12. *Nicolas Esquillan — Cinquante Ans à l'Avant-Garde de Gènie Civil*, Syndicat National du Béton Armé et des Techniques Industrialisées, Paris, 1974.

3.13. Colin Faber, *Candela — the Shell Builder*, Reinhold, New York, 1963.

3.14. Maxwell Fry and Jane Drew, *Tropical Architecture*, Batsford, London, 1964.

Chapter 4

The Concrete Code
and the Systems
of Measurement

In this chapter the codes used in reinforced concrete design are discussed and the SI metric system is briefly explained.

4.1 THE CONCRETE CODE

The Concrete Code used in this book is the *ACI Code* published in 1983 by the American Concrete Institute (Ref. 4.1); this is written in customary British/American units. There is also a metric version (Ref. 4.2), but this is not at present used in the United States. The *Commentary* on the code (Ref. 4.3) explains how the newer code rules were derived, and for some it also explains how they should be applied.

The specifications for concrete and for concrete materials are given in the *Annual Book of ASTM Standards* (Ref. 4.4).

The concrete codes of Canada, Britain, and Australia are also given in the references (Refs. 4.5 to 4.7).

4.2 SYSTEMS OF MEASUREMENT

There are three distinct systems of measurement in use at the present time. United States uses the customary British/American units, in which lengths

are measured in inches and feet and forces in pounds.   Other English-speak-
ing countries have converted to the new metric system (*Système International
d'Unités*), in which lengths are measured in millimeters and meters and forces
in newtons (Refs. 4.8 to 4.10).   Some countries still use the old metric system
in which forces are measured in kilograms or kiloponds.   There is an international agreement to adopt the SI system as soon as possible, and the United
States committed to do so, but no date has been set for a changeover.

This book is written in customary British/American units, but important
data are also given in SI units, and some of the examples are solved in both
systems of units.

4.3 THE SI METRIC SYSTEM

Using conventional American units, mass is numerically equal to weight, and
the word "weight" is frequently used when we are in fact talking about mass.
I step on a balance, and the dial swings to 140 pounds.   This is my mass.
My weight is 140 pounds-force, but we usually say just 140 pounds.   Thus
the force that gravity exerts on my body is, in conventional American units,
equal to the mass.

The same applies in the old metric system.   I step on the balance, and
the dial swings to 63.5 kilograms (kg); this is my mass, and my weight is 63.5
kilograms-force, or kiloponds.

The acceleration due to gravity at the earth's surface at sea level is 9.807
m/s^2 (meters per second squared); using 10 m/s^2 is less than 2% in error.   The
unit of force in the SI system is the newton (N), 1 N is the force due to a
mass of 1 kg that is acted on by an acceleration of 1 m/s^2.   Thus the system
is universal, and it applies on the earth's surface equally at sea level or at the
top of a high mountain; it also applies on the moon or in a spaceship.

If my mass is 63.5 kg, my weight on the earth's surface at sea level is

$$63.5 \times 9.807 = 623 \text{ N}$$

The newton is a very small unit.   It is said that Isaac Newton conceived
his laws of mechanics when he saw an apple fall to the ground.   If the apple
had a weight of 1 newton, it was a very small apple, because

$$1 \text{ newton} = 0.102 \text{ kilogram-force} = 0.225 \text{ pound-force}$$

That is, the apple had a mass of 0.225 pound, or $3\frac{1}{2}$ ounces.

In practice, forces and weights are usually measured in thousands of
newtons, or kilonewtons (kN).

$$1 \text{ kN} = 1\ 000 \text{ N} = 225 \text{ lb} = 0.225 \text{ kip}$$

In structural design we deal primarily with forces per unit area, or stresses, and the SI unit of stress is the pascal (Pa), which is the force of 1 newton acting on 1 square meter.

$$1 \text{ Pa} = 1 \text{ N/m}^2$$

The pascal is an even smaller unit, and stresses are usually measured in millions of pascals, or megapascals (MPa).

$$1 \text{ MPa} = 1 \text{ 000 kPa} = 1 \text{ 000 000 Pa} = 145 \text{ psi} = 0.145 \text{ ksi}$$

Note that

$$1 \text{ MPa} = 1 \text{ MN/m}^2 = 1 \text{ N/mm}^2$$

because $1 \text{ m} = 1 \text{ 000 mm}$ and $1 \text{ m}^2 = 1 \text{ 000 000 mm}^2$.

The SI system recognizes only conversion factors of 1 000, so that it is sometimes necessary to choose between the use of decimal fractions and of very large numbers.   It is therefore convenient to give conversion factors as follows:

	Denoted by the Prefix	Abbreviated
$10^3 \;\; = 1 \text{ 000}$	kilo	k
$10^6 \;\; = 1 \text{ 000 000}$	mega	M
$10^9 \;\; = 1 \text{ 000 000 000}$	giga	G
$10^{-3} = 0.001$	milli	m
$10^{-6} = 0.000 \text{ 001}$	micro	μ
$10^{-9} = 0.000 \text{ 000 001}$	nano	n.

Long dimensions, such as spans, are frequently given in meters, but cross-sectional dimensions and reinforcing-bar areas are always given in millimeters.   Loads are generally specified in kilonewtons (kN) or kilopascals (kPa), but stresses are stated in megapascals (MPa).   If the equation is homogeneous, it is not necessary to convert these to one set of units.   For example, a stress in MPa multiplied by one dimension in millimeters and then divided by another dimension in millimeters remains in MPa.   But if the units are mixed, it is best to convert all of them to one set of basic units.

It is convenient to choose meters and meganewtons as the basic units. Quantities in m, MN, and MPa remain unchanged.   The others are converted:

$$1 \text{ mm} = 1 \text{ m} \times 10^{-3}$$

$$1 \text{ mm}^2 = 1 \text{ m}^2 \times 10^{-6}$$

$$1 \text{ mm}^3 = 1 \text{ m}^3 \times 10^{-9}$$

$$1 \text{ kN} = 1 \text{ MN} \times 10^{-3}$$

$$1 \text{ kPa} = 1 \text{ MPa} \times 10^{-3}$$

If the result of the calculation is, say, 0.050 m, this equals 50 mm.

EXERCISES

1. Look at the *ACI Code* (Ref. 4.1), and the code *Commentary* (Ref. 4.3).   That is the code on which this book is based.   If you are not using this book in conjunction with the *ACI Code*, look at the relevant code (such as Refs. 4.5, 4.6, and 4.7).

2. If you propose to do any of the metric examples, now is a good time to practice metric conversion:

   (a) Convert (i) 227 ft; (ii) 10 ft; (iii) 5 ft, 9 in.; (iv) 8 in.; (v) $4\frac{1}{2}$ in.; (vi) $\frac{3}{4}$ in.; and (vii) $\frac{1}{2}$ in. to meters and millimeters.

   (b) Convert (i) 83 m; (ii) 3 m; (iii) 1.78 m; (iv) 300 mm; (v) 110 mm; (vi) 15 mm; and (vii) 10 mm to feet and inches.

   (c) Convert 0.15 in.^{2}/ft to mm^2/m.

   (d) Convert 300 mm^2/m to in.^{2}/ft.

   (e) Convert 250,000 lb · ft to kN · m.

   (f) Convert 2,500,000 lb · in. to kN · m.

   (g) Convert 400 kN · m to lb · ft and also to lb · in.

   (h) Convert 5,000 lb · ft/ft to kN · m/m.

   (j) Convert 20 kN · m/m to lb · ft/ft.

   (k) Convert 200 psf to kPa.

   (m) Convert 10 kPa to psf.

   (n) Convert (i) 4,000 psi and (ii) 60,000 psi to MPa.

   (p) Convert (i) 30 MPa and (ii) 400 MPa to psi.

   The answers to these subsequent numerical exercises are given in Appendix E.

REFERENCES

4.1. *Building Code Requirements for Reinforced Concrete* (*ACI 318-83*), American Concrete Institute, Box 19150 Redford Station, Detroit, 48219, 1983.

4.2. *Building Code Requirements for Reinforced Concrete* (*ACI 318M-83*), American Concrete Institute, Detroit, 1983.

4.3. *Commentary on Building Code Requirements for Reinforced Concrete* (*ACI 318-83*), American Concrete Institute, Detroit, 1983.

4.4. Part 4, *Annual Book of ASTM Standards*, American Society for Testing and Materials, Philadelphia, 1986.

4.5. *Design of Concrete Structures for Buildings* (*CAN3-A23.3-M84*), Canadian Standards Association, Rexdale, Ontario, Canada M9W 1R3, 1984.

4.6. *Structural Use of Concrete* (*BS 8110*), British Standards Institution, London, 1985 (two parts).

4.7. *SAA Concrete Structures Code* (*AS 1480-1982*), Standards Association of Australia, Sydney, 1982.

4.8. National Bureau of Standards, *The International System of Units* (*SI*), NBS Special Publication 300, U.S. Government Printing Office, Washington, D.C., 1974.

4.9. American Society for Testing and Materials, *Metric Practice Guide*, E 380-74, ASTM, Philadelphia, 1974.

4.10. Marvin H. Green, *Metric Conversion Handbook*, Chemical Publishing Co., New York, 1978.

Chapter 5

Steel and Concrete Specifications, and Control of Concrete Quality

Next, we look at the grades of steel and concrete, and the specifications governing their manufacture and use. We then examine the effect of water content on the strength of concrete and the method used for ascertaining the strength of concrete.

5.1 REINFORCING STEEL

Reinforcement for concrete consists either of individual bars or wires, or of welded fabric. Bars are easier to bend to the required shape, but fabric is convenient for slabs. There are two grades of reinforcing steel (Ref. 5.1 and Tables 5.1 and 5.2).

Grade 60 steel is more expensive than Grade 40 steel, but it costs less per unit stress, and it is therefore commonly used for reinforcement that is

TABLE 5.1   Grades of Reinforcing Steel in Accordance with ASTM 615

Grade	40	60
Minimum yield level, f_y	40,000 psi	60,000 psi
Ultimate tensile strength, f_u	60,000 psi	90,000 psi

TABLE 5.2 Metric Grades of Reinforcing Steel in
Accordance with ASTM 615 M

Grade	300	400
Minimum yield level, f_y	300 MPa	400 MPa
Ultimate tensile strength, f_u	500 MPa	600 MPa

to be stressed to the allowable limit.   On the other hand, secondary reinforcement required only for tying the primary reinforcement is more economically made from the cheaper and more easily bent Grade 40 steel.

Deformed bars are available in 11 sizes, listed in Tables 5.3 and 5.4. In addition, details of plain (i.e., not deformed) round steel bars and of deformed steel wires may be found in Part 4 of the *Annual Book of ASTM Standards* (Ref. 4.4).

Bars are produced by hot rolling, and a surface deformation is produced on the bars by an indentation in the rolls.   This ensures proper adhesion between the steel and the concrete, which is essential if the stress is to be transmitted by the concrete to the steel.   The surface of the steel must be free from grease and oil, for the same reason.   The steel may be allowed to rust slightly, as this improves the bond, but loose rust is not admissible, as this would intervene between the steel and concrete surfaces.

TABLE 5.3 Properties of Standard Metric Reinforcing Bars
in Accordance with ASTM 615

Bar Designation Number	Nominal dimensions[a]			
	Cross-sectional Area[b] (in.²)	Diameter (in.)	Perimeter (in.)	Mass (lb)
3	0.11	0.375	1.178	0.376
4	0.20	0.500	1.571	0.668
5	0.31	0.625	1.963	1.043
6	0.44	0.750	2.356	1.502
7	0.60	0.875	2.749	2.044
8	0.79	1.000	3.142	2.670
9	1.00	1.128	3.544	3.400
10	1.27	1.270	3.990	4.303
11	1.56	1.410	4.430	5.313
14	2.25	1.693	5.32	7.65
18	4.00	2.257	7.09	13.60

[a]The nominal dimensions of a deformed bar are equivalent to those of a plain round bar having the same mass per foot as the deformed bar.
[b]Tables giving the cross-sectional area of 1 to 10 bars of each size and of bars arranged at a spacing of from 4 to 18 in., are reproduced in Appendix C.

TABLE 5.4   Properties of Standard Metric Reinforcing Bars in Accordance with ASTM 615 M

Bar Designation Number	Nominal Dimensions[a]			
	Cross-sectional Area[b] (mm²)	Diameter (mm)	Perimeter (mm)	Mass (kg/m)
10	100	11.3	35.5	0.785
15	200	16.0	50.3	1.570
20	300	19.5	61.3	2.355
25	500	25.2	79.2	3.925
30	700	29.9	93.9	5.495
35	1 000	35.7	112.2	7.850
45	1 500	43.7	137.3	11.775
55	2 500	56.4	117.2	19.625

[a]The nominal dimensions of a deformed bar are equivalent to those of a plain round bar having the same mass per meter as the deformed bar.
[b]Tables giving the cross-sectional area of 1 to 10 bars of each size and of bars arranged at a spacing of from 100 to 600 mm are reproduced in Appendix C.

5.2 CEMENT

Portland cement is made by burning a mixture of finely divided limestone and clay (or shale) at a white heat at which the materials begin to fuse. After the resulting clinker has cooled, it is ground to a fine powder so as to offer a large surface to the water with which it eventually reacts.

Limestone consists of calcium carbonate, whereas clay and shale consist of aluminum silicate. In addition, a number of impurities are present in the raw materials and in the fuel used for burning the slurry, notably iron oxide. The carbon dioxide from the limestone is lost during burning.

The chemistry of portland cement is extremely complex (Ref. 5.2); however, there are four main constituents:

Tricalcium aluminate $(3CaO \cdot Al_2O_3)$
Dicalcium silicate $(2CaO \cdot SiO_2)$
Tricalcium silicate $(3CaO \cdot SiO_2)$
Tetracalcium aluminoferrite $(4CaO \cdot Al_2O_3 \cdot Fe_2O_3)$

When the cement is mixed with water, the aluminate reacts quickly, the tricalcium silicate and the aluminoferrite more gradually, and the dicalcium silicate reaction is very slow by comparison.

We can also distinguish three distinct stages in the reaction. A few hours after the cement has been brought in contact with water, the liquid paste begins to stiffen or *set*. The paste cannot be worked easily after setting has started, and the placing of the concrete must take place before setting commences.   Cement should therefore be slow setting, and a certain amount of gypsum (calcium sulfate) is added to portland cement to control setting time.

Some days after the cement has been mixed with water, it begins to *harden* and gain strength.   Since formwork cannot be removed until the concrete has sufficient strength to support its own weight, early hardening is an advantage.   The two types of portland cement (Ref. 5.3) mainly used for buildings are Type I and Type III.   Type I is the normal type, and Type III is more finely ground to offer a greater surface to the water and speed up the hardening process.   It is therefore called *high-early-strength* cement.

Although concrete gains most of its strength during the first 28 days, a gradual increase continues for many years, and this process is known as *aging*. The fact that concrete gets stronger with age, instead of losing strength, is a useful safety factor.

5.3 AGGREGATE

The normal coarse aggregate for concrete is whole gravel, crushed rock, or crushed gravel (Ref. 5.4).   The nominal maximum size is $\frac{3}{4}$ in.

The chemical composition varies from region to region depending on available materials.   Some rocks are unsuitable for use as concrete aggregates, either because they react with the cement and cause disintegration of the concrete or because they contain chemically combined water which is released during a fire and causes spalling.

Lightweight aggregate (Refs. 5.5 to 5.7) is either a natural material, such as pumice, or an artificial product.   Artificial lightweight aggregate is more expensive than normal aggregate, but it reduces the weight of the concrete, and thus the weight to be carried by the columns.   In a tall building on valuable land in the central business district, the extra rentable space made available by the reduction in column size may more than compensate for the extra cost.

Fine aggregate normally consists of sand (Ref. 5.4) either dug from a pit or collected from a beach.   When sea sand is used, it must be washed to remove all salt, because this would interfere with the chemical reaction of the cement.

5.4 ADMIXTURES

Air-entraining admixtures (Ref. 5.8), which introduce tiny air bubbles into the concrete mix, are intended to improve the workability of the mix to make it easier to get the concrete into confined spaces.   They also improve its resistance to weathering, particularly to frost.

Retarders and accelerators retard or accelerate the chemical reactions in the cement (Ref. 5.9), when more or less time is desired for the placing of the concrete, and its subsequent hardening.

Pozzolans are natural or cheap artificial materials that have natural cementing properties. Among the artificial materials are ground blast-furnace slag, and fly ash (Ref. 5.10) produced by coal-fired electric generating stations. These are waste materials that are not easily disposed of, and their use in concrete to replace a *part* of the portland cement saves money. Some pozzolans used in the correct proportion also improve the workability.

Water-reducing admixtures reduce the amount of water required for a certain workability, and because they reduce the water/cement ratio, they increase the strength of the concrete. The principal materials are modified lignosulfonates and sulfonated naphthalene or melamine formaldehyde condensates.

These materials have recently been used extensively under the name *superplasticizers* to produce high-strength concrete with normal workability and cement content. This increase in concrete strength has had an important effect in making reinforced concrete more economical for tall buildings.

All these admixtures must be used with care to ensure that the additives do not produce undesirable side effects.

Even greater care is needed when expansive cements (Ref. 5.24) are used. Concrete normally shrinks as it dries. This causes minute cracks (see Section 8.2) that are one of the undesirable features of concrete. By using a shrinkage-compensating cement, which contains a small amount of additive, the cracking can be minimized or eliminated. Shrinkage-compensating cement is constituted and proportioned such that the concrete will increase in volume after setting and during hardening. When properly restrained by reinforcement, concrete compressive stresses are induced. On subsequent drying, the shrinkage so produced, instead of causing a tensile stress to develop that might result in cracking, merely relieves the compressive stress caused by the initial expansion.

By using more potent expansive cements, self-stressing concrete is obtained (Ref. 5.11). Here the expansion, if restrained, induces compressive stresses high enough to result in a significant compression of the concrete after drying shrinkage has occurred. The high expansion in this chemically prestressed concrete (see Chapter 15) is difficult to control, and self-stressing concrete has been used only experimentally so far.

5.5 MIXING OF CONCRETE AND MIX PROPORTIONING

The *ACI Code* requires that "all concrete shall be mixed until there is a uniform distribution of materials, and shall be discharged completely before the mixer is recharged." In practice, most concrete for buildings is produced in a ready-mix plant and conforms to the appropriate standard (Ref. 5.12). Care must be taken that the concrete is placed in the mold no more than $1\frac{1}{2}$ hours after the water is added

Job-mixed concrete must conform to the same standard as ready-mixed concrete (Ref. 5.12). Mixing must continue for at least $1\frac{1}{2}$ minutes after all the materials are in the drum of the mixer. The proportion of coarse to fine aggregate is approximately 2:1, but it depends on the aggregates used, and some experimentation is necessary with unfamiliar materials.

The cement content of concrete varies from 10% to 25% of the mix. The cement is the chemically active ingredient, and high-strength concrete requires more cement. It is also the most expensive component of the concrete.

5.6 WATER

Although water is the cheapest ingredient of concrete, it is also the most important. The strength of concrete is determined by the ratio of the water content to the cement content. In 1919, Abrams (Ref. 5.13) found experimentally that the strength of concrete, other things being equal, is related to the water/cement ratio as shown in Fig. 5.1. The curve is approximately hyperbolic but it is valid only as long as the concrete is fully compacted. The

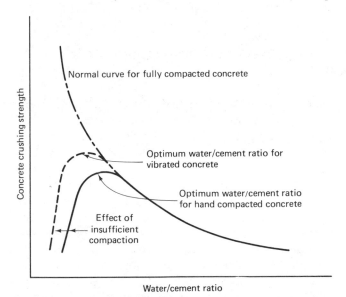

Figure 5.1 Relation between the compressive strength of concrete and its water/cement ratio. Vibration compacts the concrete, and therefore makes it possible to proceed higher up the "normal" water/cement ratio curve. Although additional water reduces the strength, it is often necessary to produce a concrete sufficiently fluid to fill the corners of the formwork and the spaces between the reinforcement.

strength can therefore be increased by vibration, which compacts concrete that would otherwise be too dry.

It is often necessary to work on the descending portion of the curve, because more water is needed than the optimum amount in order to get a concrete sufficiently fluid to penetrate between the reinforcing bars without air pockets, which would seriously reduce the strength of the structure. It is, however, important to ensure that the amount of water added is no more than the amount specified, because this lowers the strength of the concrete. Extra water makes concrete easier to place and easier to finish to a smooth surface, and supervision is often necessary to ensure that water is not improperly added. A high water/cement ratio also lowers the durability of the concrete.

The *ACI Code* in Sections 4.4 and 4.5 specifies the maximum water/cement ratio permissible in certain conditions.

The consistency of the fresh concrete (i.e., its ability to flow and be worked into the angles of the formwork and around the reinforcement) is measured by the slump test (Ref. 5.14). This is made by filling a mold made of sheet iron in the form of an open-ended truncated cone. The mold has a base diameter of 8 in., a top diameter of 4 in., and a height of 6 in. After the mold has been filled and the concrete compacted with a tamping rod, the mold is lifted vertically and the concrete cone slumps. The slump is then measured, and it is required to conform to the slump specified.

5.7 CONVEYING, DEPOSITING, AND CURING OF CONCRETE

Concrete must be conveyed from the mixer to the place of final deposit without loss of material or separation of the materials into a nonuniform mix. Interruption of concrete supply may require expensive remedial measures, such as removal of some hardened concrete. The conveying equipment must be designed to prevent that. Whenever possible, concrete should be deposited directly into the forms, as every additional handling operation adds to the cost, and it may also cause separation of the ingredients. After concreting has started, it should be carried on as one continuous operation until it is complete, or until a previously designed construction joint is reached.

Most reinforced concrete structures are designed monolithically (i.e., as if they were made from one piece of stone) but are of necessity cast in several operations. The construction joints must therefore be prepared to ensure that the assumption of monolithy is justified. This means that the joint must be cleared of accumulated dirt, debris, and water, and any soft concrete and laitance must be removed. *Laitance* is the term given to a mixture of water and cement that tends to rise to the top of concrete, particularly if it contains much water and if it has been vibrated; it gives a smooth finish to the concrete surface, but because of its high water/cement ratio, it is very weak.   The cleaned construction joint should be damp, but not wet, when the fresh concrete is poured on it.

The formwork must be designed to support the weight of liquid concrete. Concrete weighs about 144 pcf and exerts an appreciable lateral pressure, so that the strength of the formwork requires careful consideration.   Since formwork is only temporary, one does not wish to spend too much money on it. On the other hand, some serious disasters have occurred (Refs. 5.15 to 5.17) when formwork collapsed and shed liquid concrete on construction or even people below it; if the spill is large, it is usually not possible to remove the concrete before it has set.   On a less sanguine note, poorly made formwork may distort under the weight of the liquid concrete and thus produce members which are below the specified strength or which do not fit prefabricated finishes or services.

Concrete must be "cured" after it has been cast to ensure that the chemical reaction between the cement and the water occurs neither too fast nor too slow.   Concrete should therefore be kept damp and at a temperature above 50°F (10°C) for at least 7 days for normal concrete made with Type I cement, and for 3 days for high-early-strength concrete.   Accelerated curing is sometimes employed, using steam or some other source of damp heat.

In cold weather the materials may need to be preheated, and the wet concrete warmed to ensure that its temperature does not fall below 50°F. The curing of concrete also requires special attention during hot weather. Excessive temperatures may build up within the concrete due to the chemical reaction, and excessive evaporation of the mixing water may leave insufficient water for the proper hydration of the cement.   This can be remedied by spraying the concrete with cool water.

5.8 SPECIFIED COMPRESSIVE STRENGTH OF CONCRETE

The compressive strength of concrete is tested by filling a cyclindrical mold 6 in. diameter and 12 in. high with concrete, curing it under standard conditions, and testing it in compression 28 days after casting (Refs. 5.18 to 5.20). The maximum load carried by the cylinder before it crushes, divided by its cross-sectional area, must exceed the specified compressive strength f'_c of the concrete, except that a few test samples may fall below f'_c, as explained below.

A concrete mixer can mix concrete of any specified strength, but it is customary to use one of a number of standard mixes.   In conventional American units the usual strength increment is 500 psi.   The commonly used mixes for reinforced concrete are:

Grade number	3	3.5	4	4.5	5	6	7	8	9	10
f'_c (ksi)	3,000	3,500	4,000	4,500	5,000	6,000	7,000	8,000	9,000	10,000

The commonly used metric mixes are

Grade number	20	25	30	35	40	50	60	70
f'_c (MPa)	20	25	30	35	40	50	60	70

The most common concrete strengths for reinforced concrete construction are 3,500 to 6,000 psi (25 to 40 MPa). The higher-strength concretes are used mainly for prestressed concrete.

If a number of samples of a material are tested, all having nominally the same strength, some will have a slightly higher strength and others will have a slightly lower strength. In structural design it is not satisfactory to take the average strength of these samples, because the weakest piece of material might turn up at the most highly stressed part of the structure, and the structure would then fail. The specified strength f'_c is therefore the minimum strength that all test samples must exceed.

However, we cannot make this an absolute rule, because there will be a few *very* low results, however carefully the concrete is controlled. A probability of about 1 in 10 000 that the structure will not suffer any significant damage, and a probability of about 1 in 1 000 000 that it will not collapse, strike an acceptable balance between the social acceptability of a failure and the increased cost of the structure by making failure less likely. *Complete* certainty that no collapse will ever occur is unattainable.

Section 4.8.1 of the *ACI Code* lays down the frequency of testing. A sample of wet concrete must be taken at least once a day, at least once for each 150 cubic yards of concrete, and at least five samples for each class of concrete. At least two cylinders are cast from each sample, and the average of their strength constitutes one test result.

The cylinders are stored and cured under standard conditions and tested at the age of 28 days. The results are analyzed statistically (Fig. 5.2).

The *standard deviation* is computed. We have a large number of results $x_1, x_2, x_3, x_4, x_5, \ldots, x_n$. We add them up and divide by the number n of the results. This gives the *average strength* $\bar{x}$.

The standard deviation σ is a normal statistical device (Refs. 5.21 and 5.22) for assessing the extent by which individual results x depart from the average $\bar{x}$.

$$\sigma = \sqrt{\frac{\Sigma(x - \bar{x})^2}{n}} \tag{5.1}$$

This is shown in Fig. 5.2(a). Evidently, the worse the level of control, the greater the standard deviation [Fig. 5.2(b)]. If the concrete comes from a large ready-mix plant, there is a record of testing which makes it possible to assess the standard deviation likely to be achieved in the future, assuming an experienced contractor.

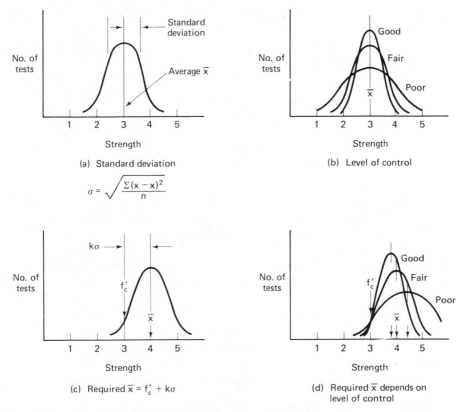

Figure 5.2 Quality control of concrete. The result of each test is plotted on the horizontal axis, and the number of results giving this strength are plotted on the vertical axis. (a) Definition of average $(\bar{x})$ and standard deviation (σ), which are normal statistical concepts. (b) Spread of test results for various levels of control. (c) Definition of specified compressive strength f'_c. If we permit a probability of 1 in 100 that an average of three consecutive strength tests fall below f'_c, we obtain from statistics that $f'_c = \bar{x} - 1.343\sigma$. (d) The average strength $\bar{x}$ depends on the level of control that we can achieve. Poor control means that a higher average strength $\bar{x}$ is required for the same f'_c.

The *specified strength* of concrete is thus defined as

$$f'_c = \bar{x} - k\sigma \qquad (5.2)$$

where $\bar{x}$ is the average strength of the concrete and σ is the standard deviation [Fig. 5.2(c)].

The report of ACI Committee 214 (Ref. 5.23) on which the 1983 *ACI Code* is based lays down three criteria:

1. A probability of less than 1 in 10 that a random individual strength test will be below that specified strength f'_c.

2. A probability of 1 in 100 that an average of three consecutive strength tests will be below the specified strength f'_c.

3. A probability of 1 in 100 that an individual strength test will be more than 500 psi below the specified strength.

Criterion 3 governs if the standard deviation is high, but for the low to moderate standard deviations obtained with good quality control, criterion 2 normally determines the required average strength $\bar{x}$ (Ref. 4.5, p. 16). This gives the condition shown in Fig. 5.2(c) for the average compressive strength $\bar{x}$ to be used in selecting the concrete proportions:

$$\bar{x} = f'_c + 1.343\sigma \tag{5.3}$$

EXERCISES

1. What is the difference between the two principal grades of reinforcing *bar*?
2. Why are reinforcing bars rolled with surface deformations?
3. What are the principal ingredients of portland cement?
4. Name some of the admixtures that are used as ingredients of concrete. What is their purpose?
5. "Water is the cheapest ingredient of concrete, but also the most important." Do you agree and, if so, why?
6. Explain the difference between the mean strength and the specified strength of a batch of concrete.
7. Explain why an appreciable amount of cement (and therefore money) can be saved by good quality control.

REFERENCES

5.1. *Standard Specification for Deformed and Plain-Billet Steel Bars for Concrete Reinforcement* (*ASTM A615-80*), published in Part 4, Annual Book of ASTM Standards, American Society for Testing and Materials, Philadelphia, 1980.

5.2. R. H. Bogue, *The Chemistry of Portland Cement*, Reinhold, New York, 1974.

5.3. *Standard Specification for Portland Cement* (*ASTM C150-78a*), published in Part 13, ASTM Standards.

5.4. *Standard Specification for Concrete Aggregates* (*ASTM C33-74a*), published in Part 14, ASTM Standards.

5.5. *Standard Specification for Lightweight Aggregates for Structural Concrete* (*ASTM C330-77*), published in Part 14, ASTM Standards.

5.6. *Lightweight Concrete* (*SP 29*), American Concrete Institute, Detroit, Mich., 1971.

5.7. *Recommended Practice for Selecting Proportions for Structural Lightweight Concrete* (*ACI 211.2-69*), American Concrete Institute, Detroit, Mich., 1969.

5.8. *Standard Specification for Air-Entraining Admixtures to Concrete (ASTM C260-77)*, published in Part 14, ASTM Standards.

5.9. *Standard Specification for Chemical Admixtures to Concrete (ASTM C494-77a)*, published in Part 14, ASTM Standards.

5.10. *Standard Specification for Fly Ash and Raw or Calcined Pozzolans for Use in Portland Cement Concrete (ASTM C618-78)*, published in Part 14, ASTM Standards.

5.11. *Klein Symposium on Expansive Cement Concretes (SP 38)*, American Concrete Institute, Detroit, 1973.

5.12. *Standard Specification for Ready-Mixed Concrete (ASTM C78a)*, published in Part 14, ASTM Standards.

5.13. D. A. Abrams, *Design of Concrete Mixes*, Bulletin No. 1, Structural Research Laboratory, Lewis Institute, Chicago, 1919.

5.14. *Standard Method of Test for Slump of Portland Cement Concrete (ASTM C143-66)*, published in Part 14, ASTM Standards.

5.15. T. H. McKaig, *Building Failures*, McGraw-Hill, New York, 1962.

5.16. J. Feld, *Lessons from Failures of Concrete Structures*, American Concrete Institute, Detroit, 1964.

5.17. J. Feld, *Construction Failure*, Wiley, New York, 1968.

5.18. *Standard Method of Making and Curing Concrete Test Specimens in the Laboratory (ASTM C192-76)*, published in Part 14, ASTM Standards.

5.19. *Standard Method of Making and Curing Concrete Test Specimens in the Field (ASTM C31-69)*, published in Part 14, ASTM Standards.

5.20. *Standard Method of Test for Compressive Strength of Cylindrical Concrete Specimens (ASTM C39-72)*, published in Part 14, ASTM Standards.

5.21. A. Huitson and J. Keen, *Essentials of Quality Control*, Heinemann, London, 1965.

5.22. J. J. Leeming, *Statistical Methods for Engineers*, Blackie, London, 1963.

5.23. *Recommended Practice for Evaluation of Compression Test Results of Field Concrete (ACI 214-65)*, American Concrete Institute, Detroit, 1965.

5.24. *Recommended Practice for Evaluation of Shrinkage-Compensating Concrete (ACI 223-77)*, American Concrete Institute, Detroit, 1977.

Chapter 6

Details of the Reinforcement

The *ACI Code* lays down rules for arranging the steel within the concrete. These are important both for economical construction and for the satisfactory performance of reinforced concrete structures. Additional reinforcement details for columns are given in Section 13.2, and for shells in Section 16.2.

6.1 CONCRETE PROTECTION FOR REINFORCEMENT

The reinforcement requires a substantial cover to ensure that the bars are surrounded by sufficient concrete to establish adequate bond between the concrete and the steel, to protect the steel from fire, and to protect the steel against rusting.

Protection against corrosion depends on the degree of exposure. Steel rusts in contact with water; the corrosion product occupies more space than the steel, and it is thus liable to burst and eventually disintegrate the concrete. Long before this happens, however, the water-soluble rust produces unsightly streaks on the concrete surface, which cannot be removed without leaving a mark. Repair of cracked concrete similarly disfigures architectural concrete.

Rusting, which is an oxidizing process, does not occur as long as the steel is surrounded by alkaline material, and hydrated cement contains free lime (CaO). This prevents rusting in spite of the ability of water to permeate concrete to a limited extent, and in spite of the formation of microcracks; the latter are the inevitable by-product of the concrete taking up its normal stress under the service loads (see Section 1.3). Thus the reinforcement does

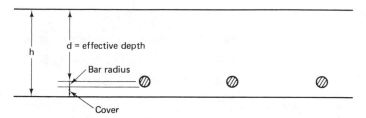

Figure 6.1   Overall depth (h), effective depth (d), and cover in reinforced concrete slabs.   The effective depth is measured to the center of the reinforcement.

not rust, provided that there is adequate cover to prevent the free lime, which is water-soluble, from being washed out.

Cover is defined as the minimum distance between the outside of the reinforcement and the surface of the concrete.   Thus in slabs (Fig. 6.1)

$$\text{overall depth} = \text{effective depth} + \text{bar radius} + \text{cover} \qquad (6.1)$$

If conduits or ducts are cast in the concrete, the full cover must be provided *over the outside* of the conduits or ducts, even if they are made of a noncorroding material.

In beams and columns the cover is measured to the outside of ties and stirrups (Fig. 6.2) so that for one layer of bars,

$$\text{overall depth} = \text{effective depth} + \text{radius of main reinforcing} \qquad (6.2)$$
$$\text{bars} + \text{diameter of tie} + \text{cover}$$

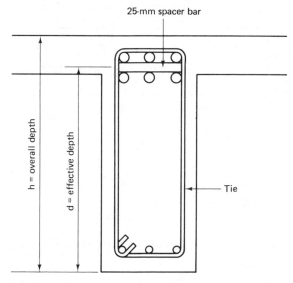

Figure 6.2   Overall depth (h), effective depth (d), and cover in reinforced concrete beams and columns.   The effective depth is measured to the center of the reinforcement.   Beams require open or closed stirrups as shear reinforcement.   In addition, beams with more than one layer of reinforcement require 1-in. spacer bars at intervals.   Columns and beams with compression reinforcement require ties to stop the buckling of the steel in compression.

and for two layers of bars,

$$\text{overall depth} = \text{effective depth} + \text{radius of spacer bar}$$
$$+ \text{ diameter of main reinforcing bar} \qquad (6.3)$$
$$+ \text{ diameter of tie} + \text{cover}$$

The *ACI Code*, Section 7.7.1, requires the following cover for cast-in-place concrete:

In concrete not exposed to the weather	
For beams and columns	$1\frac{1}{2}$ in.
For slabs, walls, and joists, provided that reinforcing bars are No. 11 or smaller (which is highly probable)	$\frac{3}{4}$ in.
For slabs, walls, and joists, provided that reinforcing bars are No. 14 or No. 18 bars	$1\frac{1}{2}$ in.
For shells and folded plate members if reinforcing bars are No. 6 or larger	$\frac{3}{4}$ in.
For shells and folded plate members if reinforcing bars are No. 5 or smaller	$\frac{1}{2}$ in.
In concrete exposed to the weather	
If reinforcing bars are No. 6 or larger	2 in.
If reinforcing bars are No. 5 or smaller	$1\frac{1}{2}$ in.
In all concrete cast directly on the earth	3 in.

The following cover is required for design in metric units:

In concrete not exposed to the weather	
For beams and columns	40 mm
For slabs, walls, and joists, provided that reinforcing bars are No. 35 or smaller (which is highly probable)	20 mm
For slabs, walls, and joists, provided that reinforcing bars are No. 45 or No. 55 bars	40 mm
For shells and folded plate members if reinforcing bars are No. 20 or larger	20 mm
For shells and folded plate members if reinforcing bars are No. 10 or No. 15 bars	15 mm
In concrete exposed to the weather	
If reinforcing bars are No. 20 or larger	50 mm
If reinforcing bars are No. 10 or No. 15	40 mm
In all concrete cast directly on the earth	75 mm

ACI Code, Section 7.7.2, permits a reduction in the cover for precast concrete members.

6.2 CONTROL OF CRACKS

Cracks are caused not merely by high stresses, but also by shrinkage of the concrete when it is restrained (see Section 8.2) and by temperature movement. They also occur in parts of the concrete structure that otherwise has low

Figure 6.3 Reinforcing detail for a
reentrant corner. Reinforcement
turning the corner must be cut and
anchored. Additional reinforcement
may be required diagonally to the
corner, at right angles to the line of a
potential crack.

stresses, but where there is a reentrant corner causing stress concentrations
(Fig. 6.3).

Reinforcement cannot prevent cracks from forming, but they can be
bridged by it to ensure that they remain small. The concrete between reinforcing bars is "unreinforced," and it is therefore appropriate in a thin concrete slab to use a large number of smaller bars closely spaced, rather than
a few bigger bars far apart. Because the use of a large number of bars adds
to the cost of handling and bar fixing and may also obstruct the placing of
the concrete, this should not be carried to extreme limits.

In a slab, reinforcement must be provided in both directions at right
angles to one another, even if structural theory requires reinforcement in one
direction only. The reinforcement in the other direction is needed to control
cracking.

Concrete cracks when the tensile strain reaches approximately 3×10^{-4}
in./in.; this ultimate strain does not vary greatly with the strength of the
concrete. The modulus of elasticity of steel is 29,000,000 psi (see Section
8.4). Consequently, the first cracks are formed in the concrete when the
steel stress reaches approximately

$$29,000,000 \times 0.000\ 3 = 8700 \text{ psi}$$

The cracks are at first very fine, and water cannot enter them because
they act like capillaries. Even when water can enter, it usually combines
with unhydrated cement over a period of time to heal the cracks.

The stress in the steel must, however, be limited to keep the cracks
small. The highest yield stress permitted by the ACI Code, Section 9.4, for
the reinforcement is $f_y = 80,000$ psi, which produces a maximum stress under
service loads (see Section 8.1) of 33,000 psi or four times the stress at which
cracks first form. Higher stresses are permitted only in prestressed concrete
(see Chapter 15).

It is advisable to use Grade 40 rather than Grade 60 steel in shells,
folded plate roofs, and other thin concrete structures exposed to the weather.
This limits the tensile steel stress under service load to 20,000 psi (as compared
to 24,000 psi for Grade 60 steel).

For structures retaining water, such as aboveground swimming pools, and for buildings subject to very severe exposure, such as sea spray or corrosive groundwater, it is wise to limit the tensile steel stress under the service loads to 14,000 psi.

6.3 MINIMUM REINFORCEMENT RATIO

Concrete contracts due to shrinkage (see Section 8.2) and due to a fall in temperature, and it expands due to a rise in temperature and also due to moisture movement in humid weather.   The movement would not set up any stress if it was not resisted, but because concrete structures are generally rigid, there is usually some restraint to the temperature and moisture movement.

A minimum amount of reinforcement is therefore needed (*ACI Code*, Section 7.12) in every concrete slab to resist temperature and shrinkage stresses. This is 0.002 0 of the gross concrete area for Grade 40 bars, 0.001 8 of the gross concrete area for Grade 60 bars.

The maximum spacing of this reinforcement is five times the slab thickness or 18 in., whichever is smaller.

For many reinforced concrete slabs these requirements for minimum reinforcement determine the amount of reinforcement, because the calculations based on strength give lower values (see Examples 12.1, 12.2, and 16.1).

The *ACI Code*, Section 10.5.1, requires that beams have positive reinforcement at least equal to that calculated from

$$\rho = \frac{200}{f_y} \tag{6.4a}$$

where ρ is the minimum value of the reinforcement ratio, defined in Section 9.1, and f_y is the yield stress of the reinforcement.   This requires 0.005 of the effective cross-sectional area for Grade 40 steel and 0.003 for Grade 60 steel.   In metric units,

$$\rho = \frac{1.4}{f_y} \tag{6.4b}$$

Beams for architectural or other reasons are sometimes made much larger in cross section than strength calculations require.   If they are then provided only with the very small percentage of reinforcement needed according to strength calculations, it could happen that the strength of the plain concrete beam is greater than the strength of the reinforced concrete beam after cracks have formed.   The development of a crack in the beam due to an overload would then be followed by instant collapse because the reinforcement is insufficient to resist the tension after the crack forms.   The minimum reinforcement requirement of Eq. (6.4) is intended to prevent a sudden collapse.

In reinforced concrete slabs an overload would be distributed laterally and a sudden failure is less likely. The *ACI Code* therefore considers that the rule for temperature and shrinkage reinforcement already stated is sufficient.

The rules for columns are considered in Section 13.2.

6.4 SPACING LIMITS FOR MAIN FLEXURAL REINFORCEMENT

There are upper and lower limits on the space allowed between bars. If bars are spaced too closely, the concrete may not be able to penetrate the space between the bars, and undetected air pockets could form (Fig. 6.4).

ACI Code, Section 7.6, requires a clear distance between bars not less than the bar diameter, d_b, or 1 in. for beams ($1.5d_b$ or $1\frac{1}{2}$ in. for columns).

When bars are placed in several layers, the bars in the upper layers must be directly above the bars in the lower layers, and the clear distance between layers must not be less than 1 in. (Fig. 6.2).

If bars are too far apart, the concrete may not have any reinforcement where a load concentration requires it. The ACI Code places an upper limit on the spacing of bars in walls and slabs of three times the wall or slab thickness, or 18 in., whichever is the smaller.

The rules for columns are considered in Section 13.2 and those for shells in Section 16.2.

6.5 BUNDLED BARS

When there is insufficient space for individual bars, two, three, or four parallel bars may be bundled together. The *ACI Code*, Section 7.6.6, requires that bundled bars be enclosed in stirrups or ties, and that individual bars terminate at different points, with at least a $40d_b$ stagger, where d_b is the diameter of an individual bar.

For the purpose of determining minimum bar spacing and cover, a

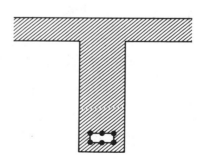

Figure 6.4 Cavity caused by reducing the bar spacing below the size of the coarse aggregate. Although the cavity may not be apparent on the surface, the bars would probably not be adequately bonded to the concrete, and the beam would then have very little strength.

bundle of bars may be treated as a single bar with a diameter that produces the same cross-sectional area as that of all the bars in the bundle.

6.6 TIES FOR COMPRESSION STEEL

Reinforcing bars are long, slender pieces of steel. When they are in compression they are liable to buckle sideways. This is easily observed by pressing with one's hand on a length of one of the thinner reinforcing bars. Compression reinforcement in beams and columns must therefore be restrained by ties (Fig. 6.2).

Ties are generally made from the smallest reinforcing bars, and their maximum spacing must not exceed:

> 16 longitudinal bar diameters
> 48 tie diameters
> The least dimension of the cross section

6.7 TERMINATION OF REINFORCING BARS

Bars are terminated when they are no longer required, or when they get as long as can conveniently be handled. It is necessary to transfer the force in the terminated bar to the concrete through surface bond between the steel and the concrete before the bar can be cut off. The maximum force in the bar occurs at the point of maximum bending moment (see Fig. 6.8 and Sections 7.3 and 9.2), and at that point the steel has its maximum stress. If the beam or slab has been economically designed, this is very close to the maximum admissible stress, f_y, that is

$$f_s = f_y$$

A development length ℓ_d is needed to transfer this force from the steel to the concrete (Fig. 6.5). The surface area of the bar that is required to make the transfer is directly proportional to the bar diameter d_b and the development length. The force in the bar is proportional to the cross-sec-

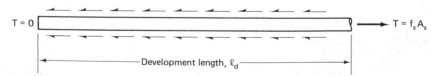

Figure 6.5 The tensile force in the bar $T = f_s A_s$ (where $f_s \leq f_y$ is the maximum stress in the bar) is transferred to the concrete by bond, that is, by friction between the surface of the bar and the concrete, over a development length ℓ_d.

tional area, and hence to d_b^2. Therefore, the larger the bar diameter, the greater must be its development length.

If the liquid concrete is cast to a depth of more than 12 in., settlement of the concrete below the top bars may reduce the effectiveness of the bond. Thus top bars require a greater development length. A *top bar* is defined as one that has 12 in. or more of concrete cast below it. In practice, bars at the top of reinforced concrete slabs, which are rarely sufficiently deep, are, "other than top bars," but the bars at the top of reinforced concrete beams, which are usually deeper, are top bars.

The formulas for the development length are given in Sections 12.2 and 12.3 of the *ACI Code*, and they are explained in more advanced books on reinforced concrete design (e.g., Ref. 7.5). The values obtained from these formulas are listed in Tables 6.1 and 6.2.

A structural member is not always long enough to accommodate the development length in tension. In that case a 180° or 90° hook (a bar bent through a right angle) is used (Fig. 6.6). The anchorage value of hooks is shown in Table 6.3. However, hooks produce stress concentrations within the concrete, and if too many hooks converge on the same part of the concrete structure, they can cause failure. We therefore rely on developing the stress in a straight bar whenever there is enough room (Fig. 6.5), and use hooks only where they are essential. They are usually needed at the end of a beam or slab. If the number of hooks becomes excessive, it is better to redesign the structure. Hooks are not useful on compression reinforcement.

TABLE 6.1 Development Length ℓ_d for Grade 60 Steel and Grade 4 Concrete[a] in Inches

Bar No.	Tension Reinforcement				Compression Reinforcement
	Bar Spacing < 6 in.		Bar Spacing ≥ 6 in.		
	Top Bar	Other Than Top Bar	Top Bar	Other Than Top Bar	
3	12.60	12.00	12.00	12.00	7.11
4	16.80	12.00	13.44	12.00	9.49
5	21.00	15.00	16.80	12.00	11.86
6	25.20	18.00	20.18	14.40	14.23
7	31.88	22.77	25.50	18.22	16.60
8	41.97	29.98	33.58	23.98	18.97
9	53.13	37.95	42.50	30.36	21.40
10	67.48	48.20	53.98	38.58	24.10
11	82.88	59.20	66.30	47.36	26.75
14	119.5	85.39	95.63	68.31	32.12
18	212.5	151.8	170.0	121.4	42.82

[a]For other steel and concrete grades, see formulas in Sections 12.2 and 12.3 of the *ACI Code* (Ref. 4.1) or Table B3, Ref. 7.5.

TABLE 6.2 Development Length ℓ_d for Grade 400 Steel and Grade 25 Concrete[a] in Millimeters

| | Tension Reinforcement | | | | |
| | Bar Spacing < 150 mm | | Bar Spacing ≥ 150 mm | | |
Bar No.	Top Bar	Other Than Top Bar	Top Bar	Other Than Top Bar	Compression Reinforcement
10	380	300	304	300	226
15	538	384	430	307	320
20	672	480	538	384	390
25	1120	800	896	640	504
30	1568	1120	1254	896	598
35	2240	1600	1792	1280	714
45	3360	2400	2686	1920	874
55	5600	4000	4480	3200	1128

[a]For other steel and concrete grades, see formulas in Sections 12.2 and 12.3 of the *Metric ACI Code* (Ref. 4.2) or Table B3, Ref. 7.5.

TABLE 6.3 Anchorage Value of Standard Hook, Expressed as an Equivalent Development Length ℓ_d.[a]

For Grade 60 Steel and Grade 4 Concrete (in.)[b]		For Grade 400 Steel and Grade 25 Concrete (mm)[b]	
Bar No.	Equivalent Development Length	Bar No.	Equivalent Development Length
3	7.11	10	226
4	9.49	15	320
5	11.86	20	390
6	14.23	25	504
7	16.60	30	598
8	18.97	35	714
9	21.40	45	874
10	24.10	55	1128
11	26.75		
14	32.12		
18	42.82		

[a]If a bar terminates in a standard hook, this equivalent length can be subtracted from the straight length ℓ_d of straight tension reinforcement given in Tables 6.1 and 6.2.

[b]For other steel and concrete grades, see formulas in Section 12.5 of the *ACI Code* (Refs. 4.1 and 4.2) or Table B.4, Ref. 7.5. Hooks do not make a useful contribution to the development of strength in compression reinforcement, and compression steel should not be terminated with a hook.

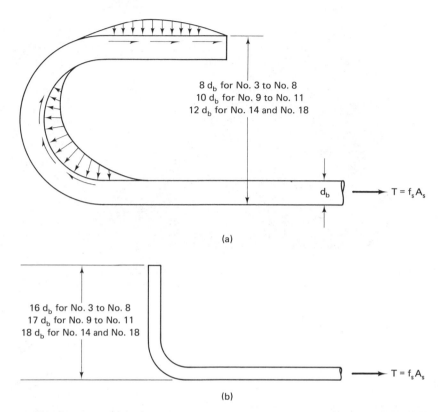

Figure 6.6   A 180° hook (a) provides anchorage for a reinforcing bar in tension by setting up compressive stresses in the concrete as shown.   These increase the friction between the steel and the concrete.   The action of a 90° hook (b) is similar. Hooks are not useful for anchoring bars in compression.

Because hooks cause stress concentrations in the concrete (Fig. 6.7) and additional stresses in the steel, Section 7.2 of the *ACI Code* lays down minimum bend diameters (Fig. 6.6).   The hooks bent in accordance with these requirements take up an appreciable amount of space for the larger bar diameters.   Thus the width of a standard hook for a No. 18 bar, measured from the outside to the outside of the bar, is $12d_b = 27$ in. (Fig. 6.6).

The *ACI Code* also requires that the bar be continued for an additional length ℓ_a beyond the point where it is no longer required, that is, beyond the point of inflection, where the bending moment changes from positive to negative, or vice versa (Fig. 6.8), or beyond the point where the bending moment is so reduced that some of the bars can be cut off.   The bar must be continued beyond this theoretical cutoff point for the additional distance

$$\ell_a \geq \text{effective depth of beam or slab} \tag{6.5}$$
$$\geq 12d_b \quad \text{(diameter of bar to be cut off)}$$

whichever is larger (Fig. 6.9).

Figure 6.7 Spalling of concrete over a 180° hook due to inadequate cover. Note the crushing of the concrete inside the hook. A similar result is observed on a 90° hook with inadequate cover.

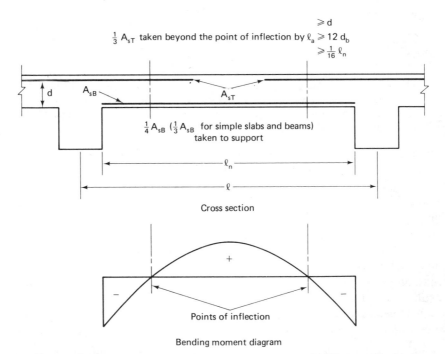

Figure 6.8 A part of the positive (A_{sB}) reinforcement and of the negative (A_{sT}) reinforcement must be taken beyond the point of inflection where the bending moment changes sign.

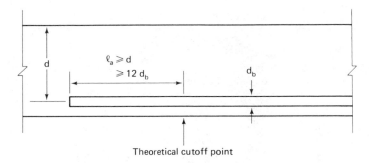

Figure 6.9 Additional length ℓ_a required beyond the theoretical cutoff point.

Although the positive steel can be cut off or bent up when the bending moment changes sign at the point of inflection, in practice the need for adequate anchorage requires that a proportion of the reinforcement be taken beyond the point of inflection to the support (*ACI Code*, Sections 12.12 and 12.13).

Thus at least one-fourth of the positive reinforcement (Fig. 6.8) in continuous slabs (one-third in simple slabs) must be taken to the face of the support.   The same rule applies to beams, except that the reinforcement must be taken 6 in. into the support.

At least one-third of the negative reinforcement must have an additional length

$$\ell_a \geq \text{effective depth of beam or slab}$$

$$\geq 12d_b \tag{6.6}$$

$$\geq 1/16 \quad \text{clear span of beam or slab}$$

whichever is the largest (Fig. 6.8).

At the end of a beam or slab, this anchorage length is not available for the negative steel, and a hook is used instead (see Figs. 9.3 and 9.4).

6.8 SPLICING OF REINFORCEMENT

In steel structures two pieces of steel are joined by welding or bolting, but it is much cheaper to join reinforcing bars by bonding them to the concrete and transferring stress through the concrete to another bar.   The bars run parallel to one another for a splice length that depends on the development length ℓ_d.   It is specified in Section 12.16 of the *ACI Code*.

This is the most common method; but in heavily reinforced structural members there may not be enough room for the doubling of the bars which a splice requires.   It is then necessary to join the reinforcing bars by welding,

TABLE 6.4 Part of Bar Schedule

Mark	Grade	Size Bar No.	Number Required	Length	Straight	Bend Type (refer to bending schedule)[a]	Center-to-Center Spacing (in.)
G 1	60	8	10	11'3"	X		8
G 2	60	6	8	2'6"		D	6
G 3	60	3	23	4'0"	X		16
.	.	.	.	.	.	.	
.	.	.	.	.	.	.	

[a]The bending schedule is not reproduced.

which is permitted by the *ACI Code*, Section 12.15.3, provided that the welded joint develops the full strength of the bar, that is, the welding does not lower the strength of the steel.   Mechanical connections between bars, for example by screw thread and nut, are also permitted, but these are more expensive than welding, and welding is more expensive than a splice through bond.

6.9 REINFORCED CONCRETE DRAWINGS AND SCHEDULES

Before a reinforced concrete structure can be built, it is necessary to schedule all the reinforcing bars required for it.   Each bar is identified by a mark. When the bars are cut and bent, if required, they are labeled with this mark, which identifies their location in the structure.

The schedule lists the bar mark, the grade of the steel, the size of the bar, the number of bars of this mark required, their length, and details of their bends if any (Table 6.4).   The bends are sometimes listed in the same schedule, or reference is made to a separate bar bending schedule.

It would not be practicable to draw every bar in a reinforced concrete slab, nor is it necessary.   It is generally sufficient to indicate the width over which bars of each mark are to be used (e.g., Fig. 12.9).   The bar mark number then refers to the bar schedule, which gives the details of the bar.

Bar schedules and sometimes drawings are now commonly made by computer for all but the smallest reinforced concrete structures.

EXERCISES

1. Why do reinforcing bars require cover, whereas structural steel can, in certain circumstances, be used without protection?
2. How much cover is needed:
   (a) In 'suspended' floor slabs?

(b) In floor slabs cast directly on the earth?

(c) In roof slabs?

(d) In beams exposed on the facade of the building?

(e) In interior columns?

3. Why does concrete require a minimum amount of reinforcement, and how much is it?

4. What is the minimum space required between reinforcing bars?

5. What is the maximum space permitted between reinforcing bars in slabs?

6. Explain the requirements for anchoring a reinforcing bar:

   (a) By means of a 180° hook.

   (b) By means of a 90° hook.

   (c) By means of a straight development length.

7. Why is it necessary to continue some reinforcing bars beyond the point where they are no longer required for resistance to the bending moment?

8. Why do we normally splice reinforcing bars by bond with the concrete, instead of using welded or screwed joints, as in structural steel?

Chapter 7

The Design
of the Reinforced
Concrete Structure

We examine the moments and forces produced in structures by the loads. The stresses due to these moments and forces are considered in Chapter 8.

7.1 LOADS

The loads that reinforced concrete (and other) structures are required to support are set out in other codes, for example *Building Code Requirements for Minimum Design Loads in Building and Other Structures (ANSI A58.1)* of the American National Standards Institute.

We distinguish between two kinds of vertical loads. Those due to the weight of the building are called *dead loads*. They include the weight of the reinforced concrete structure, which is sometimes greater than all the other loads put together. The weight of reinforced concrete depends on the amount of reinforcement, which weighs more than three times as much as concrete; but we usually take the weight of reinforced concrete, irrespective of the amount of steel, as 144 pcf* (2 400 kg/m³ or 24 kN/m³). Other dead loads include the weight of walls and permanent partitions, finishes, and any part of the building that is not normally removed.

---

* 144 pcf equals 12 lb per square foot per inch thickness, so a 6-in. slab weighs 72 psf.

Vertical live loads consist of the people in the building, furniture, movable partitions, and anything that could be removed by the occupants without requiring the aid of a construction team.

The *horizontal loads* are the forces due to wind and earthquakes. Wind exerts pressure on the windward face of the building and suction on the other faces. The roof, unless it is a steeply sloping roof, is also subject to suction.

Earthquake forces occur only in some places where there are faults in the earth's crust (Fig. 1.3). The earth moves suddenly, but the building does not immediately move with it because of its inertia. This movement of the earth relative to the building exerts an inclined force, which can be resolved into horizontal and vertical components. Usually buildings have adequate strength to resist the vertical forces, but in earthquake zones the horizontal forces may be greater than those due to wind (see Sections 10.7, 11.6, and 13.11).

7.2 ANALYSIS OF THE STRUCTURE

The behavior of the structure under the action of these loads is then analyzed. The structure must have an adequate foundation (see Chapter 14), and any differential statement that may occur must be included in the design of the frame.

Although the *ACI Code* is frequently called an ultimate-strength code, this refers only to the design of the sections. The moments and forces in the structural frame are determined by elastic analysis, using the actual, or *service loads* that the structure carries.

This mixture of elastic and ultimate-strength design may at first seem incongruous, but structural design is an empirical technology rather than an exact science, and the mixture of elastic and ultimate-strength methods is not new in reinforced concrete design. The *ACI Code* before 1971 used elastic analysis both for determining the forces and moments in the frame, and for determining the stresses at sections subject to bending, but the design of columns was based on ultimate strength. In the new code the design of all sections is based on ultimate strength, but the elastic analysis remains for the frame.

We do not use ultimate-strength methods for determining the forces and moments in the structure (except optionally in slabs by the yield-line theory, see Section 7.5) because our research on the ultimate strength of concrete structures, as distinct from individual concrete members, is not sufficiently far advanced. We do use ultimate strength for the concrete sections, because it saves time and material.

TABLE 7.1(a) Bending Moment Coefficients for the Dead Loads on Continuous Beams and Slabs[a]

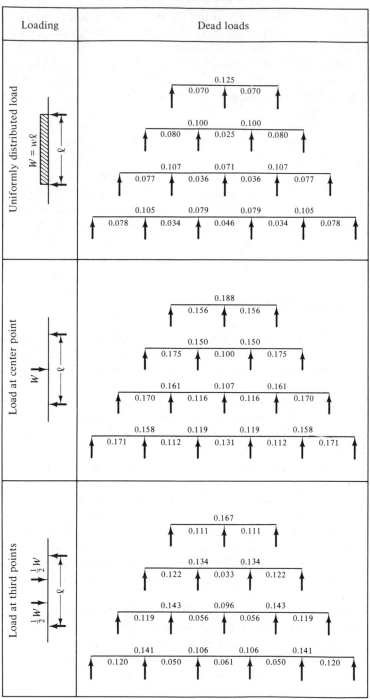

[a]It is assumed that all spans are fully loaded, M = coefficient $\times$ $W\ell$.

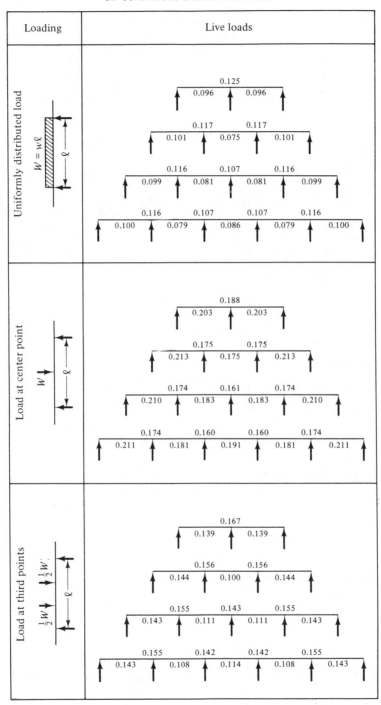

[a]It is assumed that either alternate or adjacent spans are fully loaded, whichever gives the highest bending moment (see Fig. 7.1). M = coefficient $\times$ $W\ell$.

TABLE 7.2 Maximum Positive and Negative Bending Moment in Fixed-End Beams and Slabs

Types of Loading	Maximum Negative Bending Moment	Maximum Positive Bending Moment
Point load W at midspan ($\frac{1}{2}\ell$, $\frac{1}{2}\ell$)	$-\frac{1}{8}W\ell$	$+\frac{1}{8}W\ell$
Two loads $\frac{1}{2}W$ at third points ($\frac{1}{3}\ell$, $\frac{1}{3}\ell$, $\frac{1}{3}\ell$)	$-\frac{1}{9}W\ell$	$+\frac{1}{18}W\ell$
Three loads $\frac{1}{3}W$ ($\frac{1}{4}\ell$ spacing)	$-\frac{5}{48}W\ell$	$+\frac{3}{48}W\ell$
Point load W at $a\ell$ from A, span ℓ (A to B)	$M_A = -a(1-a)^2 \; W\ell$ $M_B = -a^2(1-a) \; W\ell$	$+2a^2(1-a)^2 \; W\ell$
Total load W uniformly distributed over span ℓ	$-\frac{1}{12}W\ell$	$+\frac{1}{24}W\ell$
Total load W triangular (peak at midspan), $\frac{1}{2}\ell$, $\frac{1}{2}\ell$	$-\frac{5}{48}W\ell$	$+\frac{3}{48}W\ell$
Total load W triangular over span ℓ (A to B)	$M_A = -\frac{1}{10}W\ell$ $M_B = -\frac{1}{15}W\ell$	$+0.043\,W\ell$

TABLE 7.3 Maximum Bending Moments in Simply Supported Beams and Slabs

Types of Loading	Maximum (Positive) Bending Moment
	$+\frac{1}{4}W\ell$
	$+\frac{1}{6}W\ell$
	$+\frac{1}{6}W\ell$
	$+a(1-a)W\ell$
	$+\frac{1}{8}W\ell$
	$+\frac{1}{6}W\ell$
	$+0.128W\ell$

7.3 ELASTIC ANALYSIS

A digital computer is always employed for the elastic analysis of large reinforced concrete frames, by the matrix-displacement or other appropriate method,
and readers are referred to one of the many textbooks on computer methods
(e.g., Ref. 7.1).   For smaller frames the moment distribution method is
sometimes more convenient.

If the stiffness of the floor structure is appreciably higher than the
stiffness of the supporting columns, the floor may be considered continuous
over the supports, and continuous beam theory used.   Table 7.1 gives the
solution for beams of two, three, four, or five equal spans.

If the ends of a slab are rigidly restrained, the theory of fixed-end beams
may be used.   Table 7.2 gives the results for fixed-end beams for various
types of loading.   If the ends of a slab are supported on load-bearing brick
or block walls that do not restrain the ends, as often happens in small buildings,
the slab may be considered simply supported.   Table 7.3 gives the maximum
bending moments for simply supported beams with various kinds of loading.

The various elastic methods of design are described in many textbooks
on structural design (e.g., Ref. 7.2) and in some specialized textbooks (e.g.,
Ref. 7.3).   The elastic design of slabs spanning in two directions, including
flat slabs and flat plates, is considered in Chapter 12, and that of shells and
folded plates in Chapter 16.

The *ACI Code*, Chapter 8, gives some rules for elastic analysis:

1. The modulus of elasticity of concrete (in psi) is given by the empirical
formula

$$E_c = \gamma_c^{1.5} \times 33 \sqrt{f_c'} \tag{7.1}$$

where γ_c is the density of the concrete, pcf; and f_c' is the specified concrete
strength, psi (see Section 5.8).   For normal-density concrete this becomes

$$E_c = 57,000 \sqrt{f_c'} \tag{7.2}$$

2. The stiffness of the section may be calculated by any reasonable
assumption, but this shall be used consistently throughout the whole analysis.
The most common practice is to treat the concrete section as uncracked for
this purpose, and ignore the reinforcement.

3. The effective span ℓ of a member in a continuous frame is the distance
between the centerlines of the supporting members.   In the analysis of a
member not built integrally with its supports, ℓ is the center-to-center distance
between supports or $\ell_n + h$, whichever is the lesser (where ℓ_n is the clear
span from support face to support face, and h is the overall depth of the
member).

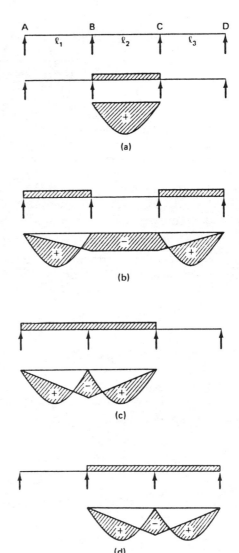

Figure 7.1 Alternative loading conditions for the live load. (1) Alternate spans loaded: this produces (a) the maximum positive moment in span ℓ_2 and (b) the maximum positive moments in spans ℓ_1 and ℓ_3, because the negative moments M_B and M_C have their minimum values when the span ℓ_2 is unloaded. (2) Adjacent spans loaded: this produces (c) the maximum negative moment M_B, and (d) the maximum negative moment M_C.

4. In considering vertical live loads it is permissible to consider only two alternative loading conditions (Fig. 7.1):

  (a) Alternate spans loaded, and all other spans unloaded

  (b) Adjacent spans loaded and all other spans unloaded

5. The negative bending moments at the supports may be increased or decreased by not more than

$$m = 0.20\left(1 - \frac{\rho - \rho'}{\rho_b}\right)$$

The terms ρ, ρ', and ρ_b are defined in Appendix B.

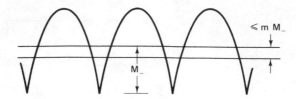

Figure 7.2 The negative bending moments at the supports may be increased or decreased by not more than $m M_-$ provided that the base line of the entire bending moment diagram is displaced to maintain equilibrium.

In a normal beam or slab this amounts to about 15%. The *ACI Code* in Section 8.4.3 lays down some limiting conditions. When this adjustment is made, the base line of the entire bending moment diagram is displaced to maintain the conditions of equilibrium (Fig. 7.2).

6. A rectangular frame may be analyzed as a series of parallel bents, each bent consisting of a floor and the columns above and below (Fig. 7.3). The columns are assumed fixed at their ends for the purpose of moment distribution.

7.4 BENDING MOMENT COEFFICIENTS

The *ACI Code*, Section 8.3.3, gives coefficients for bending moments (Fig. 7.4) which can be used without any analysis of the structure, provided that spans are approximately equal (i.e., they do not differ by more than 20% of

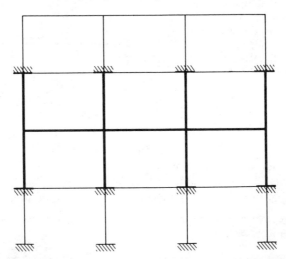

Figure 7.3 Analysis of individual bents of rigid frames. Two systems of these bents are commonly considered, spanning at right angles to one another. Thus each part of the floor structure is part of two bents at right angles to one another.

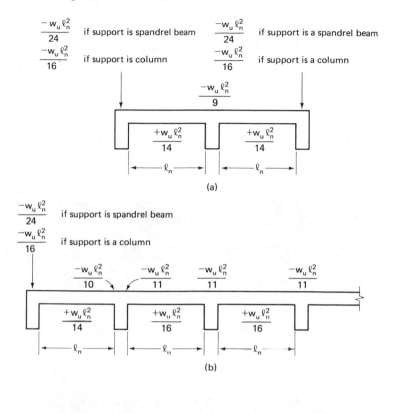

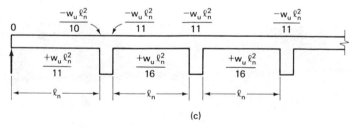

Figure 7.4   Bending moment coefficients according to the *ACI Code*, Section 8.3.3. (a) For two-span members rigidly joined to their end supports. (b) For members of more than two spans rigidly joined to their end supports. (c) For members of more than two spans unrestrained at their end supports.

the shorter), there are two or more spans, and the live load does not exceed three times the dead load.

These coefficients are empirical adjustments of the bending moment coefficients obtained from the theory of continuous and of fixed-end beams and slabs. They are convenient for minor structures which do not require a more accurate analysis.

7.5 PLASTIC ANALYSIS OF REINFORCED
CONCRETE STRUCTURES

Steel structures, particularly if they are not too tall, can be designed according to the assumption that plastic hinges form as the steel yields, until the structure is turned into a mechanism and collapses (Ref. 7.4).

Plastic hinges are also formed by reinforced concrete if the percentage of the reinforcement is small, so that the steel yields before the concrete is crushed. This is the basis of the ultimate strength method for the design of

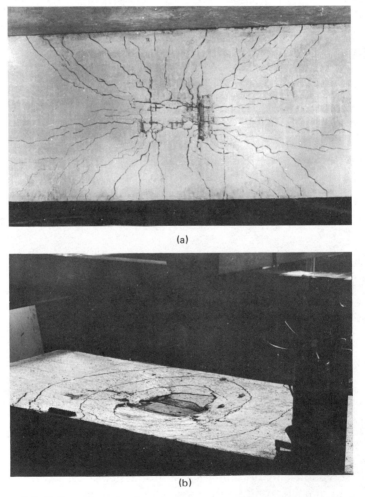

(a)

(b)

Figure 7.5 Failure of rectangular concrete two-way slab, simply supported along the edges, carrying a concentrated central load. (a) Tension face. (b) Compression face.

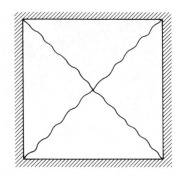

Figure 7.6 Yield-line theory for a square reinforced concrete slab, carrying a uniformly distributed load. The plastic resistance moment along the cracks of the slab equals the moment due to the loads acting on the slab.

reinforced concrete sections (see Chapter 8). In order to design an entire frame by plastic analysis, however, one would need to be certain that all the hinges form before the concrete disintegrates. Plastic analysis (also called limit design) of reinforced concrete frames is therefore possible only in exceptional circumstances. The problem is discussed by Ferguson (Ref. 7.5, Chap. 11).

The *ACI Code*, however, takes account of redistribution of moments due to plasticity. The moment adjustment described in Section 7.3 and Fig. 7.2 is based on the plastic deformation of the steel.

Furthermore, plastic hinges form in reinforced concrete slabs (Fig. 7.5) before the concrete disintegrates, and these cracks can be idealized as shown in Fig. 7.6. The four parts of the slab then form a mechanism, and the plastic resistance moment of the slab along the cracks can be equated to the moment due to the ultimate loads acting on the slab. This *yield-line theory* was developed by K. W. Johansen in Denmark in 1942 (Ref. 7.6). It is specially permitted by the Scandinavian countries and Australia, and it can be used in the United States under Section 13.3.1 of the *ACI Code*; this is specifically stated in the *ACI Code Commentary* (Ref. 4.5, p. 83). The method is fully explained by Ferguson (Ref. 7.5, Chap. 13).

7.6 EXPERIMENTAL STRESS ANALYSIS

Some structures or structural members are of a shape so complex that it is difficult to express their geometry in terms that are amenable to theoretical analysis. An approximate theoretical solution can be made by analyzing a simpler but slightly different structure. An answer can also be obtained by making an exact reduced-scale model and measuring the stresses in it by means of strain gauges (Fig. 7.7). Usually, both methods are used in conjunction.

Experimental stress analysis has been used for a number of imaginative, but complicated, reinforced concrete structures (Ref. 7.7). However, its

Figure 7.7 Model of a tall reinforced concrete building (the 64-story MLC Center in Sydney), built from Plexiglas to a scale of 1:95, and tested at the University of Sydney Architectural Science Laboratories of the University of Sydney (structural design: Civil and Civic Pty Ltd; test: Dr. J. S. Gero). The wind load is simulated by horizontal wires loaded by dead weights. The strains in the model are measured with electric-resistance strain gauges, and its deflections are measured with dial gauges. The stresses and deflections in the actual structure are deduced from these measurements by dimensional theory (Ref. 7.7), and compared with the results of a computer analysis of the structure.

scope is decreasing because many structures of complex or undefined geometry can now be modeled mathematically by computer.

EXERCISES

1. How do you calculate the dead load due to the weight of the reinforced concrete structure?
2. What is a service load?
3. What is the difference between elastic design and ultimate-strength design?

4. Briefly describe the methods used for calculating the bending moments and shear forces in reinforced concrete structures.
5. Does the modulus of elasticity vary with the grade of reinforcing steel? What is its value?
6. Does the modulus of elasticity vary with the grade of concrete? What is its value for 3,000-psi concrete?
7. Calculate the dead load due to a reinforced concrete slab 5 in. thick.
8. Calculate the load due to a reinforced concrete slab 125 mm thick.

REFERENCES

7.1. H. B. Harrison, *Computer Methods in Structural Analysis*, Prentice-Hall, Englewood Cliffs, N.J., 1973.
7.2. H. J. Cowan, *Architectural Structures*, American Elsevier, New York, 1976.
7.3. F. B. Bull and G. Sved, *Moment Distribution in Theory and Practice*, Pergamon Press, Elmsford, N.Y., 1964.
7.4. L. S. Beedle, *Plastic Design of Steel Frames*, Wiley, New York, 1958.
7.5. P. M. Ferguson, *Reinforced Concrete Fundamentals*, 4th ed. (in conventional American units), Wiley, New York, 1979. SI edition, Wiley, New York, 1981.
7.6. K. W. Johansen, *Yield-Line Theory*, Cement and Concrete Association, London, 1963 (originally published in Danish in 1943).
7.7. H. J. Cowan and others, *Models in Architecture*, Elsevier, London, 1968.

Chapter 8

Basic Assumptions Made in the Design of Reinforced Concrete Sections

This book is really all about the design of reinforced sections, and we now look at the basic assumptions. This chapter contains many of the design constants that you will need later in the book, and it explains how they are obtained.

8.1 DEVELOPMENT OF THE STRAIGHT-LINE THEORY

It is an historical fact that ultimate strength design preceded elastic design (Ref. 2.6). In the seventeenth and eighteenth centuries the strength of structural parts was determined by load tests to destruction, usually employing cannon balls whose weight was known with comparative accuracy (Fig. 8.1). The *theory* of ultimate strength design, however, dates only from the early twentieth century, whereas the theory of elastic design was developed in the late eighteenth and early nineteenth century and had reached a considerable degree of sophistication when in 1826 L. M. H. Navier published *Résumé des leçons . . .*, which established the trend of structural design for a century to come. It was therefore natural that Koenen in 1886 (see Section 2.4) and Coignet and Tedesco in 1894 framed their theories of reinforced concrete design in elastic terms. The latter is still in use; it is the straight-line theory of Appendix B of the *ACI Code*.

Figure 8.1 Full-scale bending test by direct loading. (From Peter Barlow, *An Essay on the Strength and Stress of Timber*, printed for A. J. Taylor at the Architecture Library, London, 1817, Pl. V.)

Section B.3 of the *ACI Code* gives the maximum permissible stresses under the action of the service loads.   For bending these are:

0.45 f'_c for concrete in compression, where f'_c is the specified compressive strength of concrete (see Section 5.8)

20,000 psi for Grade 40 steel

24,000 psi for Grade 60 steel

The ratio 0.45 gives a factor of safety of 1/0.45 = 2.2, and the maximum permissible stress for Grade 60 steel gives a factor of safety of 60/24 = 2.5. These safety factors take the place of the load factors and strength-reduction factors described in Section 8.6.

The behavior of the structure under the action of the service loads is then analyzed, assuming that steel and concrete deform elastically and that the maximum permissible stresses must not be exceeded.

The straight-line theory will not be considered further in this book, except in Chapters 15 and 16, which deal with prestressed concrete and concrete shells.   It is described in detail in any textbook on reinforced concrete design published before 1971 (e.g., Ref. 8.5) and in some contemporary books (e.g., Ref. 7.5, App. A, pp. 669–698).

For normal reinforced concrete design the straight-line theory has been superseded by the ultimate strength method, described in Sections 8.3 to 8.6.

8.2 SHRINKAGE AND CREEP OF CONCRETE

Like timber, but unlike steel, concrete deforms with moisture movement. This does not, in itself, cause any stresses, but when the deformation is restrained, stresses are set up. The combination of concrete, which deforms with moisture movement, and steel, which does not, therefore creates stresses in both materials.

There are two types of moisture movement in concrete: shrinkage, which is caused by contraction of the concrete as it loses some of its water, partly by chemical reaction and partly by evaporation; and creep, which is caused by water being squeezed from the fine pores of the concrete under sustained load. Since both are caused by the water and the cement, the aggregate being inert, shrinkage and creep both increase with the water content and also with the cement content.

The shrinkage of concrete ranges from 2×10^{-4} to 5×10^{-4} in./in. It can be measured if a precise figure is required, but it is generally sufficiently accurate to take it as 3×10^{-4}.

When a piece of concrete of unit length ℓ_1 (Fig. 8.2) is reinforced, there is a combination of two materials: the reinforcement, which would, if acting separately, remain at the length ℓ_1, and the concrete, which acting separately would shrink by $s\ell_1$ to the length ℓ_2. Acting in combination, the length becomes ℓ_3. This means that the steel is given a compressive strain ϵ_s, and the concrete a tensile strain ϵ_c. The compressive strain in the steel causes a slight redistribution of stress, but the tensile strain in the concrete can be more than its ultimate tensile strain, in which case the concrete cracks.

This is not a problem in the direction in which a slab or wall is bent. Reinforcement is provided on one face to resist the bending moment, and the other face is in compression. This eliminates the tensile strain. However, if there is no bending moment at right angles to the first direction, the re-

Figure 8.2   The shrinkage of reinforced concrete. The concrete would, if unrestrained by the reinforcement, shrink by $s\ell_1$ from the original unit length ℓ_1 to ℓ_2. Because it is restrained by the steel, the length becomes ℓ_3. This produces a compressive steel strain ϵ_s and a tensile concrete strain ϵ_c.

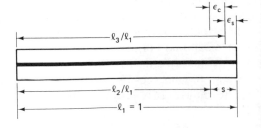

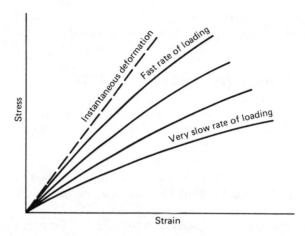

Figure 8.3 Stress–strain diagram of concrete in compression for various rates of loading. Because of creep, the strain increases with time, and the diagram becomes more curved. The effective modulus of elasticity of concrete is reduced with time.

inforccment is theoretically not required, and shrinkage cracking is then likely. A similar problem arises from movement due to tempcrature changes. The *ACI Code* therefore provides that a minimum amount of *temperature and shrinkage reinforcement* must be provided in any direction (see Section 6.3).

Shrinkage may also set up more substantial stresses, for example, in rigid arches, where it alters the shape of the arch.

Creep is caused by sustained loads. The weight of the concrete itself is a large part of the load it carries, and there may be other large dead loads. Thus the instant elastic deformation of concrete under load continues to increase over a period of time (Fig. 8.3).

The structure deflects instantly under load. During its lifetime this amount is gradually increased by about a further 200%. *ACI Code*, Section 9.5.2.5, therefore requires that the immediate elastic deflection caused by the sustained loads (i.e., the dead loads and long-action live loads) be multiplied by

$$\left(2 - 1.2\frac{A_s'}{A_s}\right) \geq 0.6$$

where A_s' is the cross-sectional area of the compression reinforcement, if any, and A_s that of the tension reinforcement. If there is no compression reinforcement, the factor is 2, and the elastic flection is therefore trebled to allow for the effect of creep.

The initial elastic deflection can be allowed for, but the creep deflection may cause jamming of doors and windows and cracking of brittle finishes or partitions. To avoid this, it is essential to limit the initial elastic deflection (see Section 8.7), particularly in flexible structures such as flat plates (see Section 12.5), since the creep deflection is proportional to the elastic deflection.

8.3 STRESS–STRAIN DIAGRAM OF CONCRETE

When reinforced concrete became an important structural material toward the end of the nineteenth century, precise methods of strain measurement were already well developed, and many of the earliest experimenters observed that the concrete stress–strain diagram was not straight, but curved. Several theories based on curved stress blocks were proposed even before 1910, but these were not ultimate strength theories, but rather elastic theories with curved stress blocks.

The observation of creep in concrete (Fig. 8.3) seemed to provide an explanation of the curvature of the stress–strain diagram, but the ultimate strain of concrete is in fact higher than could be explained by creep alone. In the 1940s and subsequently it was shown (Refs. 8.1 and 8.2) that the stress–strain diagram has a descending portion if the load is gradually reduced after the maximum has been reached, and ultimate strains as high as 3×10^{-3} were observed. These are very much smaller than the ultimate strains of steel, but also much higher than had been thought likely before.

The deformation of concrete at high loads appears to be due to the formation of microcracks (Ref. 8.3) too fine to be observed by the naked eye, but detectable with ultrasonic equipment (Ref. 8.4). This is quite unlike the plastic deformation of steel, and it is for this reason that the plastic design methods that have been developed for rigid steel frames (commonly known as limit design) can be adapted to reinforced concrete only with caution (see Section 7.5).

Although some very complicated mathematical equations have been devised to fit the experimental stress–strain diagram of concrete (Ref. 2.11), these are not really required, as long as we know the area contained under the stress–strain diagram of Fig. 8.4 and the location of the centroid of this area.

The *ACI Code*, Section 10.2.6, states that the relationship between concrete compressive stress distribution and concrete strain may be assumed to be rectangular, trapezoidal, parabolic, or any other shape that results in a prediction of strength in substantial agreement with the results of comprehensive tests. In practice, the rectangular stress block is almost invariably used because it produces the simplest equations.

Section 10.2.7 of the *ACI Code* defines the equivalent rectangular stress block as shown in Fig. 8.5. It has the same area as the curved stress block that it replaces, and it has its centroid at the same distance from the right-hand edge of the diagram. It is therefore fully equivalent to the curved stress block. The application of this rectangular stress block to design is discussed in Section 9.1.

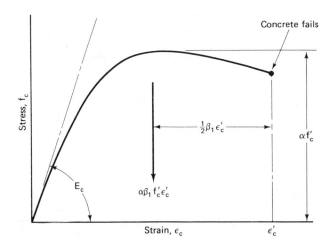

Figure 8.4 General shape of the concrete stress–strain diagram in compression. f'_c = specified compressive strength of the concrete required by the designer, as determined from standard cylinder tests. α = a factor that converts the cylinder crushing strength to the compressive strength observed when the same concrete is tested to destruction in a beam; the *ACI Code* assumes that α = 0.85. β_1 = a factor locating the centroid of the area under the stress–strain diagram. E_c = modulus of elasticity of the concrete under short-term loading. ϵ'_c = ultimate strain of concrete. It has been found (Ref. 2.11) that the ultimate strain varies but little with the characteristic strength of the concrete, f'_c, because the stronger concretes are more brittle; the *ACI Code* requires that ε'_c = 0.003, irrespective of the value of f'_c.

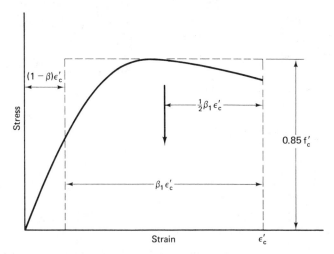

Figure 8.5 Equivalent rectangular stress block for concrete. The rectangular stress block is intended to have the same area as the curved stress block, and to have its centroid at the same distance from the right edge of the diagram (i.e., ½ $\beta_1\epsilon'_c$). The value of β_1 = 0.85 for $f'_c \leq 4,000$ psi.

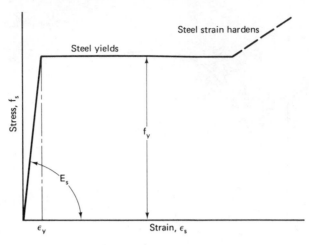

Figure 8.6 Idealized stress–strain diagram of structural-grade reinforcing steel. f_y = yield stress; ϵ_y = yield strain; E_s = modulus of elasticity = 29,000,000 psi.

8.4 STRESS–STRAIN DIAGRAM OF STEEL

Steel deforms elastically, with a modulus of elasticity E_s, until the yield point f_y is reached (Fig. 8.6). The steel then yields, or deforms plastically, at the constant stress f_y. The end of the plastic stage corresponds to a strain much greater than the ultimate concrete strain, and although strain hardening of the steel can occur, it is neglected.

Cold-worked bars and hard-drawn wire do not show a pronounced yield point, but the stress–strain diagram is idealized for the purpose of ultimate-strength design into a perfectly elastic range with a modulus of elasticity E_s, followed by a perfectly plastic stage during which the steel deforms at a constant stress f_y.

The *ACI Concrete Code* specifies a modulus of elasticity E_s = 29,000,000 psi for all types of steel.

8.5 BASIC ASSUMPTIONS MADE
IN THE ULTIMATE-STRENGTH THEORY

The two most important assumptions made by the *ACI Code* are evidently to specify the shape of the stress–strain diagrams for steel and concrete to be used in the calculations, already discussed in Sections 8.3 and 8.4.

In addition, the *ACI Code* assumes:

1. *Strains in the concrete and the steel are directly proportional to the distance from the neutral axis* (Fig. 8.7). This means that sections plane before bending remain plane after bending. It is the normal assumption made in

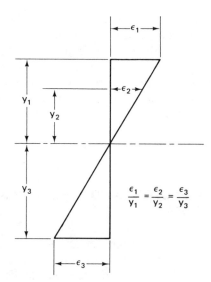

Figure 8.7 Strains in the concrete and the steel are directly proportional to the distance from the neutral axis.

$$\frac{\epsilon_1}{y_1} = \frac{\epsilon_2}{y_2} = \frac{\epsilon_3}{y_3}$$

elastic theories of bending (see Section 8.1), and it may at first sight seem surprising that this should apply also to the heavily cracked sections at the ultimate load. However, extensive experimental observations have shown this to be correct. It is assumed, by implication, that there is perfect bond between the steel and the surrounding concrete. If the steel is at a depth y_3 below the neutral axis, then ϵ_3 is the strain both of the steel and of the concrete at y_3.

2. *The tensile strength of the concrete may be neglected.* At very low loads, before the concrete cracks [Fig. 8.8(b)], the concrete can be analyzed

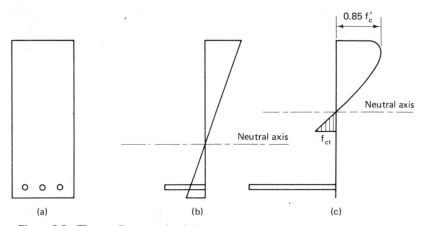

Figure 8.8 The tensile strength of the concrete may be neglected in reinforced concrete flexural members. (a) Dimensions of the section. (b) Stress distribution at low loads before the concrete cracks. (c) Stress distribution at failure.

by the theory of composite sections.   After the concrete cracks in tension, only the part that is still stressed below the ultimate tensile concrete strength f_{ct} contributes to the resistance moment.   If $f_c'/f_{ct} = 12$ (it usually lies between 10 and 20), then the ratio of the area of the compressive stress block to the shaded tensile concrete stress block is about 150:1, and the ratio of the moments of both blocks about the neutral axis is about 1 700:1.   The neglect of the shaded tensile concrete stress block thus produces an error of less than 0.1%, which is negligible.   Thus the entire tensile resistance is provided by the reinforcing steel.

8.6 LOAD FACTORS AND STRENGTH-REDUCTION FACTORS

Ultimate-strength design determines the load at which a structural member would actually fail.   Since we do not want it to fail, we use in our calculations *ultimate loads U*, which are the actual loads carried by the structure multiplied by a *load factor*.   The load factor provides a margin of safety to allow for the possibility that the specified loads may be exceeded.   We know the dead loads with considerable accuracy, and they require a lower factor than the live loads, which are more likely to be exceeded with a possible change in the function of the building.   Thus the first alternative condition considered by Section 9.2 of the *ACI Code* is

$$U = 1.4D + 1.7L \tag{8.1}$$

In this equation D is the force or moment caused by a dead load, L is the force or moment caused by a live load, and U is the force or moment corresponding to the ultimate load condition.

The other two equations are concerned with wind and earthquakes (if any).   The stresses caused by the dead and the live loads normally increase the stresses caused by wind or earthquakes, in which case the ultimate force or moment

$$U = 0.75(1.4D + 1.7L + 1.7W) \tag{8.2}$$

where W is the force or moment due to either wind loads or earthquake forces.

On the other hand, wind, in particular, might cause an uplift, in which case the dead loads and the live loads should be taken at their lowest possible value; the lowest possible effect of a live load is zero, when it is not acting at all.   Then

$$U = 0.9D + 1.3W \tag{8.3}$$

In addition to the load factors, the *ACI Code* also prescribes reduction factors, ϕ, to allow for simplifications in the structural analysis, variations in the quality of the materials (see Chapter 5), and inaccuracies in the formwork or the placement of the reinforcement which produce variations in the cross-sectional dimensions.   Values for these ϕ factors are given in Section 9.3 of the *ACI Code* as follows:

Flexure and axial tension 0.90

Shear and torsion 0.85

Spirally reinforced columns 0.75

Tied columns 0.70

Bearing 0.70

The value of ϕ is highest for bending and axial tension, which depend on the strength of the steel, since this is less variable than the strength of the concrete.   As will be shown in Section 9.4, the *ACI Code* ensures that reinforced concrete beams always fail in tension.

The value of ϕ is reduced for shear and torsion because of our still inadequate understanding of the mechanics of failure of reinforced concrete in shear and torsion.

The lower values of ϕ for columns are for compression failure, where the strength of concrete is critical; a higher value of ϕ is applied to columns that fail in tension (see Section 13.8).   In addition, a column failure is more disastrous than a beam failure, since the collapse of a column also causes the failure of the beams and slabs carried on all floors above it.

No allowance is made in these factors for the type of occupancy, because this is already made in the loading code; nor is there any allowance for large arithmetic errors.

Although arithmetic *errors* are inadmissible, a high degree of accuracy is not required in reinforced concrete calculations.   It is sufficient to take calculations to three significant figures and be certain of the accuracy of two significant figures in the final result.

8.7 CONTROL OF DEFLECTION

Excessive deflection can limit the serviceability of a structural member for three main reasons:

1. If the structural member forms part of a flat roof, it may interfere with the drainage of the roof.   If it forms part of a floor that needs to be perfectly level (e.g., in a bowling alley), it may interfere with the function of the floor.
2. If the deflection is visible to the eye, it may cause concern about the safety of the structure.   Sagging beams and slabs worry people, even when they are perfectly safe.
3. The deflection of a member, although perfectly safe, may cause failure

of finishes applied to the member, or may damage other members in contact with it. For example, plaster ceilings are cracked by excessive deflections. Nonstructural brick or block partitions that touch the underside of a reinforced concrete slab may be damaged, and doors or windows in the partitions may jam because of excessive deflection of the slab. This can be avoided by leaving a small gap between the top of the partition and the slab, but this is not always practicable.

Point 3 is the most important reason for limiting deflection.

The general formula for the deflection of a beam or slab carrying a uniform distributed load w over a span ℓ is

$$\Delta = \frac{kw\ell^4}{EI} \tag{8.4}$$

where E is the modulus of elasticity [see eq. (7.1)], I is the moment of inertia, and k is a constant. The value of I must be calculated in accordance with the provisions of Section 9.5.2 of the *ACI Code*, which considers the effect of the reinforcement and of cracks in the concrete on the stiffness of the beam.

The *ACI Code* in Table 9.5(b) lays down the maximum permissible computed deflections. The immediate elastic deflection due to a service live load L is limited to $\ell/180$ for flat roofs and to $\ell/360$ for floors.

The part of the total deflection occurring after nonstructural finishes have been attached is also limited. It includes the immediate elastic deflection due to the live load, and in addition, the long-term deflection due to all sustained loads. The long-term deflection includes the effect of creep (see Section 8.2); that is, if there is no compression reinforcement, the elastic deflection in Table 8.1 is trebled for loads added after the nonstructural elements were attached, and doubled for loads that were already in position. The *ACI Code* distinguishes between roofs and floors that support or have

TABLE 8.1 Calculation of Elastic Deflection, $\Delta = kw\ell^4/EI$[a]

Conditions of Support	k
Cantilever, i.e., one end free, the other rigidly restrained	$0.1250 = 1/8$
Simply supported beam or slab, i.e., both ends simply supported	$0.0130 = 5/384 = 1/77$ approx.
Propped cantilever, i.e., one end simply supported, the other rigidly restrained	$0.0054 = 1/185$ approx.
Built-in beam or slab, i.e., both ends rigidly restrained	$0.0026 = 1/384$

[a] w = uniformly distributed load; ℓ = span; E = modulus of elasticity; I = moment of inertia.

TABLE 8.2 Minimum Thickness for Beams and One-Way Slabs

Beam or Slab	Solid One-way Slab	Beam or Ribbed One-way Slab
Is simply supported	$\ell/20$	$\ell/16$
Has one end continuous	$\ell/24$	$\ell/18.5$
Has both ends continuous	$\ell/28$	$\ell21$
Is cantilevered	$\ell/10$	$\ell/8$

TABLE 8.3 Minimum Thickness for Two-Way Slabs, Flat Slabs, and Flat Plates

Type of Slab (see Chapter 12)	Minimum Thickness	
	(in.)	(mm)
Slabs with beams on all four edges	$3\frac{1}{2}$	90
Slabs without beams, but with drop panels	4	100
Slabs without beams or drop panels	5	120

attached to them nonstructural elements likely to be damaged by large deflections, and those unlikely to be damaged by large deflections. For the first category the deflection is limited to $\ell/480$, and for the second to $\ell/240$.

If an elastic frame analysis is made (see Section 7.3) using a digital computer, maximum deflections can be included in the printout. If the calculations are made by hand with a calculator, the work involved in determining deflections is appreciable.

For small and simple structures it is sufficient to use the alternative method given in Sections 9.5.2.1 and 9.5.3.1 of the *ACI Code*, which lays down minimum thicknesses for beams and slabs. For beams and one-way slabs these are as given in Table 8.2. The table gives the minimum thickness h as a ratio of the span l for beams and slabs that do not support or are attached. to partitions or other construction likely to be damaged by large deflections. It is to be used when deflections have not been computed. The ratios apply only to normal-weight concrete and Grade 60 steel. Correction factors are given in the *ACI Code* for lightweight concrete or Grade 40 steel.

For two-way slabs, flat slabs and flat plates, the minimum thickness h is the value obtained from Eq. (8.5)* or the dimension given in Table 8.3, whichever is larger.

$$h = \frac{\ell_n}{36\,000}\left(800 + \frac{f_y}{200}\right) \tag{8.5}$$

*This is Eq. (9.13) of the *ACI Code*. The code states two additional equations (9.11 and 9.12) that are more complicated. These give a lower minimum thickness for some slabs.

where ℓ_n is the length of the clear span in the long direction and f_y is the yield stress of the steel in psi.

EXERCISES

1. What is shrinkage? Why is it likely to produce cracking in reinforced concrete structures?
2. What is creep? What is its effect on the stress distribution in reinforced concrete structures?
3. Why does creep increase the deflection of reinforced concrete beams and slabs? Roughly, by how much?
4. What is a load factor? Why is it higher for live loads than for dead loads?
5. What is a strength-reduction factor? Why is it higher for beams than for columns?
6. Why is the ultimate strain of steel so much greater than that of concrete?
7. What is the straight-line theory? For what purpose is it used?
8. What is the ultimate-strength theory? How does it differ from the straight-line theory?
9. What are the basic assumptions made in the ultimate-strength theory?
10. What is the cause of the additional long-term deflection that occurs in reinforced concrete structures? Why is it more difficult to counter than short-term deflection?

REFERENCES

8.1. D. Ramaley and D. McHenry, *Stress–Strain Curves for Concrete Strained beyond the Ultimate Load*, U.S. Department of the Interior, Bureau of Reclamation, Laboratory Report No. SP-12, Denver, Colo., 1947.

8.2. H. J. Cowan, "Inelastic Deformation of Reinforced Concrete in Relation to Ultimate Strength," *Engineering*, Vol. 174, 1952, pp. 276–278.

8.3. H. J. Cowan, "Representation of the Inelastic Deformation of Concrete by Means of a Mechanical Model," *Nature*, Vol. 178, 1956, pp. 278–279.

8.4. R. Jones, "A Method of Studying the Formation of Cracks in Materials under Stress," *British Journal of Applied Physics*, Vol. 3, 1952, pp. 229–232.

8.5. C. W. Dunham, *The Theory and Practice of Reinforced Concrete*, McGraw-Hill, New York, 1953.

Chapter 9

Ultimate-Strength Design of Reinforced Concrete Slabs and Rectangular Beams

After eight chapters of introduction, we now get down to business. This chapter contains the basic theory of reinforced concrete design and nine examples.

9.1 ANALYSIS OF RECTANGULAR SECTIONS: TENSION FAILURE

We now apply the assumptions outlined in Chapter 8 to the design of rectangular sections, the simplest and the most common type of concrete section since it includes concrete slabs.

Let us consider a rectangular section of width b (in the case of a slab this is a unit width of 1 ft), which is reinforced with an area of tension steel A_s at an effective depth d (i.e., the center of the steel is at a depth d below the top of the section). We do not consider the overall depth of the section since the concrete below the steel is assumed to be cracked and thus does not contribute to the strength of the section; the overall depth is, of course, required for construction.

At the service load the section is loaded by a bending movement M which produces only elastic strains [Fig. 9.1(b)], varying from ε_c at the top of the section to ε_s at the center of the steel. We now increase the bending moment M until the steel strain reaches the yield strain ε_y [Fig. 9.1(c)]. Since

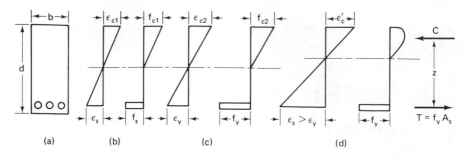

Figure 9.1 Primary tension failure in bending. (a) Dimensions of section. (b) Elastic stage. (c) Steel yields. (d) Failure. ε_y, f_y = yield strain, yield stress of steel; ε_c' = ultimate strain of concrete.

we are dealing with the case of primary tension failure, we have a section in which the steel has just reached its yield strength f_y while the concrete has not yet reached its ultimate strength f_c'.

We now have a tensile force in the section, formed by the area of tension steel and the stress in it, which is the yield stress of the steel:

$$T = f_y A_s \tag{9.1}$$

This force can be increased no further. As the steel strain increases to

$$\varepsilon_s > \varepsilon_y$$

the steel stress remains constant at the yield stress f_y (see Fig. 8.6).

A further increase in the moment M can be produced only by lengthening the moment arm z. The centroid of the compressive stress block therefore moves toward the compression face, and this in turn means that the neutral axis must rise, with a consequent rapid increase in the concrete strain. When the concrete strain reaches $\varepsilon_c' = 0.003$ (see Section 8.3), the concrete reaches its crushing strength, and the crushing of the concrete produces the collapse of the reinforced concrete member. When this happens, the ultimate resistance moment of the section, M', has been reached [Fig. 9.1(d)].

Let us now determine the value of the moment M', using the equivalent rectangular stress block of Fig. 8.5.

We have a resultant tensile force of

$$T = f_y A_s \tag{9.1}$$

and a resultant compressive force (Fig. 9.2) of

$$C = 0.85 f_c' b a$$

where the stress $0.85f_c'$ is drawn as the horizontal dimension and a as the vertical dimension of the equivalent rectangular stress block, and b is the width of the section (1 ft for a slab).

We chose the rectangular stress block so that it has the same area as

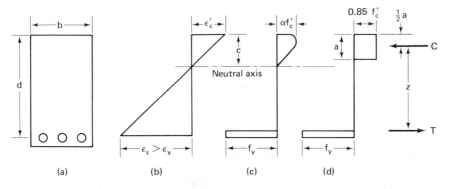

Figure 9.2 Solution of primary tension failure by means of the equivalent rectangular stress block. (a) Dimensions of section. (b) Strain distribution a failure. (c) Stress distribution at failure. (d) Stress distribution, using the equivalent rectangular stress block.

the curved stress block, and it has its centroid at the same depth as the curved stress block (i.e., the line of action of the resultant compressive force C is not affected). This is achieved by making

$$a = \beta_1 c$$

where c is the depth of the neutral axis at the ultimate moment M'. The value of β_1 is specified as the *ACI Code*, Section 10.2.7 (see Section 8.3 of this book). Thus the resultant compressive force is

$$C = 0.85f_c'b\beta_1 c \qquad (9.2)$$

The equations become much simpler if we introduce the reinforcement ratio

$$\rho = \frac{A_s}{bd}$$

which makes

$$T = f_y\rho bd \qquad (9.3)$$

For horizontal equilibrium the resultant tensile and compressive forces must balance:

$$C = T$$

$$0.85f_c'b\beta_1 c = f_y\rho bd$$

which gives the neutral-axis depth ratio,

$$\frac{c}{d} = \frac{f_y\rho}{0.85f_c'\beta_1} \qquad (9.4)$$

The resultant compressive force C acts at the centroid of the equivalent rectangular stress block (i.e., at a depth $\frac{1}{2}a$), and the length of the moment arm is thus

$$z = d - \tfrac{1}{2}a = d - \tfrac{1}{2}\beta_1 c \tag{9.5}$$

and the moment arm ratio [from Eq. (9.4)]

$$\frac{z}{d} = 1 - \frac{\tfrac{1}{2}\beta_1 c}{d} = 1 - \frac{1}{2}\frac{f_y}{0.85 f_c'}\rho = 1 - \frac{0.588 \rho f_y}{f_c'} \tag{9.6}$$

It is sufficiently accurate to use

$$\frac{z}{d} = 1 - \frac{0.6 \rho f_y}{f_c'} \tag{9.7}$$

The values of z/d obtained from this equation are listed in Tables 9.1 and 9.2.

The ultimate resistance moment is obtained by taking moments about the line of action of the resultant compressive force C, which gives Tz. We must now include the strength-reduction factor ϕ to comply with the requirements of the *ACI Code*, which gives

$$M' = \phi Tz = \phi f_y A_s z = \phi f_y \rho \frac{z}{d} bd^2 \tag{9.8}$$

TABLE 9.1 Variation of Moment Arm with Reinforcement Ratio:
$z/d = 1 - 0.6\,\rho f_y/f'_c$

Reinforcement Ratio, ρ	$f_y = 40\ 000$ psi				$f_y = 60\ 000$ psi			
	$f'_c = 3000$	4000 (psi)	5000	6000	$f'_c = 3000$	4000 (psi)	5000	6000
0.001	0.99	0.99	0.99	0.99	0.99	0.99	0.99	0.99
0.002	0.98	0.99	0.99	0.99	0.98	0.98	0.99	0.99
0.003	0.98	0.98	0.99	0.99	0.96	0.97	0.98	0.98
0.004	0.97	0.98	0.98	0.98	0.95	0.96	0.97	0.98
0.006	0.95	0.96	0.97	0.98	0.93	0.94	0.96	0.96
0.008	0.94	0.95	0.96	0.97	0.90	0.93	0.94	0.95
0.010	0.92	0.94	0.95	0.96	0.88	0.91	0.93	0.94
0.012	0.90	0.93	0.94	0.95	0.86	0.89	0.91	0.93
0.014	0.89	0.92	0.93	0.94	0.83	0.87	0.90	0.92
0.016	0.87	0.90	0.92	0.94	0.81	0.86	0.88	0.90
0.018	0.86	0.89	0.91	0.93	a	0.84	0.87	0.89
0.020	0.84	0.88	0.90	0.92	a	0.82	0.85	0.88
0.025	0.80	0.85	0.88	0.90	a	a	0.82	0.85
0.030	a	0.82	0.86	0.88	a	a	a	0.82

[a]Outside the range of applicability of this formula; see Section 9.4.

TABLE 9.2 Variation of Moment Arm Ratio with Reinforcement Ratio:
$z/d = 1 - 0.6\rho f_y/f'_c$.

Reinforcement Ratio, ρ	$f_y = 300$ MPa				$f_y = 400$ MPa			
	$f'_c = 20$	25 (MPa)	30	35	$f'_c = 20$	25 (MPa)	30	35
0.001	0.99	0.99	0.99	0.99	0.99	0.99	0.99	0.99
0.002	0.98	0.99	0.99	0.99	0.98	0.98	0.98	0.99
0.003	0.97	0.98	0.98	0.98	0.96	0.97	0.98	0.98
0.004	0.96	0.97	0.98	0.98	0.95	0.96	0.97	0.97
0.006	0.95	0.96	0.96	0.97	0.93	0.94	0.95	0.96
0.008	0.93	0.94	0.95	0.96	0.90	0.92	0.93	0.94
0.010	0.91	0.93	0.94	0.95	0.88	0.90	0.92	0.93
0.012	0.89	0.92	0.93	0.94	0.86	0.88	0.90	0.92
0.014	0.87	0.90	0.92	0.93	0.83	0.87	0.89	0.90
0.016	0.86	0.88	0.90	0.92	0.81	0.85	0.87	0.89
0.018	0.84	0.87	0.89	0.91	a	0.83	0.86	0.88
0.020	0.82	0.86	0.88	0.90	a	0.81	0.84	0.86
0.025	a	0.82	0.85	0.87	a	a	a	0.83
0.030	a	0.78	0.82	0.85	a	a	a	a

[a]Outside the range of the applicability of this formula; see Section 9.4.

The bending moment M_u due to the loads [each multiplied by the appropriate load factor; see Eqs. (8.1) to (8.3)] must be less; that is,

$$M_u \leq M'$$

if the structure is to be safe.

9.2 SIMPLE SLAB PROBLEMS

Example 9.1A

Determine the effective depth and the main reinforcement required for a one-way reinforced concrete slab continuous over two equal clear spans of 13 ft 6 in. to carry a dead load of 100 psf and a live load of 60 psf, using Grade 3.5 concrete, Grade 60 steel, and a reinforcement ratio of 0.005.

Data given: $D = 100$ psf, $L = 60$ psf, $\ell_n = 13.5$ ft, $b = 1$ ft, $f'_c = 3,500$ psi, $f_y = 60,000$ psi, $\phi = 0.90$, and $\rho = 0.005$.

Bending moment: From Eq. (8.1) the load factor for the dead load is 1.4 and the load factor for the live loads is 1.7. Therefore, the ultimate load is

$$w_u = 1.4 \times 100 + 1.7 \times 60 = 242 \text{ psf}$$

Using the approximate bending moment coefficients of the *ACI Code* (see Fig. 7.4), the maximum bending moment is

$$M_u = \frac{-w_u\ell_n^2}{9} = \frac{-242 \times 13.5^2}{9} = -4{,}900 \text{ lb} \cdot \text{ft/ft}$$

that is, 4,900 pounds-feet per foot width of slab.

Effective depth: The resistance moment of the reinforced concrete slab, from Eq. (9.8), is

$$M' = \phi f_y \rho \frac{z}{d} bd^2$$

The strength-reduction factor $\phi = 0.90$ for flexure (Section 8.6). The yield stress of the steel, $f_y = 60{,}000$ psi; the unit width of the slab $b = 1$ ft ($= 12$ in.). From Table 9.1, $z/d = 0.95$ for $\rho = 0.005$, $f_y = 60{,}000$ psi, and $f'_c = 3{,}500$ psi. Putting $M' = M_u$, and ignoring the minus sign, which indicates that the reinforcement is required on the top of the slab, yields

$$4{,}900 = 0.90 \times 60{,}000 \times 0.005 \times 0.95 \times 1 \times d^2$$

This gives the effective depth of

$$d = 4.370 \text{ in.}$$

Main reinforcement: The cross-sectional area of the reinforcement required for a unit width (1 ft $= 12$ in.) of the slab is

$$A_s = \rho bd = 0.005 \times 12 \times 4.37 = 0.262 \text{ in.}^2/\text{ft.}$$

From Table 5.3, a No. 4 bar has a cross-sectional area of 0.20 in.^{2}, so we need 1.31 bars per foot, or a spacing of $12/1.31 = 9.16$ in. The nearest round figure is 9 in. Alternatively, this can be read directly from Table C.3 in Appendix C. *Use No. 4 bars on 9-in. centers* ($=0.27$ in.^{2}/ft).

Overall depth: The overall depth required, from Eq. (6.1), is

$$h = d + \text{bar radius} + \text{cover} = 4.37 + 0.25 + 0.75 = 5.37 \text{ in.}$$

This is increased to the nearest preferred dimension. In conventional U.S. units the depth of slabs increases in ½-in. increments, or sometimes in 1-in. increments, so $h = 5\frac{1}{2}$ in.

Example 9.1B

Determine the effective depth and the main reinforcement required for a one-way reinforced concrete slab continuous over two equal clear spans of 4 m, to carry a dead load of 5 kPa and a live load of 3 kPa, using Grade 25 concrete, Grade 400 steel, and a reinforcement ratio of 0.005.

Data given: $D = 5$ kPa, $L = 3$ kPa, $\ell_n = 4$ m, $b = 1$ m, $f'_c = 25$ MPa, $f_y = 400$ MPa, $\phi = 0.90$, and $\rho = 0.005$.

Bending moment: From Eq. (8.1) the load factor for the dead load is 1.4 and the load factor for the live loads is 1.7. Therefore, the ultimate load is

$$w_u = 1.4 \times 5 + 1.7 \times 3 = 12.1 \text{ kPa}$$

Using the approximate bending moment coefficients of the *ACI Code* (see Fig. 7.4), the maximum bending moment is

$$M_u = \frac{-w_u \ell_n^2}{9} = \frac{-12.1 \times 4^2}{9} = -21.51 \text{ kN} \cdot \text{m/m}$$

that is, 21.51 kilonewton meters per meter width of slab. Since the maximum stresses are in megapascals, we convert this to meganewtons.

$$M_u = -21.51 \times 10^{-3} \text{ MN} \cdot \text{m/m}$$

Effective depth: The resistance moment of the reinforced concrete slab, from Eq. (9.8), is

$$M' = \phi f_y \rho \frac{z}{d} bd^2$$

The strength-reduction factor $\phi = 0.90$ for flexure (Section 8.6). The yield stress of the steel, $f_y = 400$ MPa, the unit width of the slab $b = 1$ m. From Table 9.2, $z/d = 0.95$ for $\rho = 0.005$, $f_y = 400$ MPa, and $f'_c = 25$ MPa. Putting $M' = M_u$, and ignoring the minus sign, which indicates that the reinforcement is required on the top of the slab, yields

$$21.51 \times 10^{-3} = 0.90 \times 400 \times 0.005 \times 0.95 \times 1 \times d^2$$

This gives the effective depth of

$$d = 0.112 \text{ m} = 112 \text{ mm}$$

Main reinforcement: The cross-sectional area of the reinforcement required for a unit width of the slab is

$$A_s = \rho bd = 0.005 \times 1\,000 \times 112 = 560 \text{ mm}^2/\text{m}$$

From Table 5.4, the smallest reinforcing bar, No. 10, has a cross-sectional area of 100 mm^2, so that we need 5.6 bars per meter, or a spacing of $1\,000/5.6 = 179$ mm. The nearest round figure is 175 mm. Alternatively, this can be read directly from Table C.4 in Appendix C. *Use No. 10 bars on 175-mm centers* ($= 571$ mm^2/m).

Overall depth: The overall depth required, from Eq. (6.1), is

$$h = d + \text{bar radius} + \text{cover} = 112 + 5 + 20 = 137 \text{ mm}$$

This is increased to the nearest preferred dimension, so that $h = 150$ mm.

Example 9.2

Complete the design of the slab of Example 9.1.

Data given: $D = 100$ psf, $L = 60$ psf, $\ell_n = 13.5$ ft, $b = 1$ ft, $f'_c = 3,500$ psi, $f_y = 60,000$ psi, $\phi = 0.90$, and $\rho = 0.005$.

Apart from the reinforced concrete theory explained in Example 9.1, there are some other considerations that require attention. We now complete the design of the slab, including the reinforced concrete calculations at a more rapid rate.

Bending moment: The ultimate load is

$$w_u = 1.4 \times 100 + 1.7 \times 60 = 242 \text{ psf}$$

and the maximum bending moment (from Fig. 7.4) is

$$M_u = \frac{-w_u \ell_n^2}{9} = \frac{-242 \times 13.5^2}{9} = 4,900 \text{ lb} \cdot \text{ft/ft}$$

Effective depth: The resistance moment of the slab is

$$M' = \phi f_y \frac{z}{d} b d^2$$

From Table 9.1, $z/d = 0.95$:

$$4,900 = 0.90 \times 60,000 \times 0.005 \times 0.95 \times 1 \times d^2$$

$$d = 4.37 \text{ in.}$$

Main reinforcement:

$$A_s = \rho b d = 0.005 \times 12 \times 4.37 = 0.262 \text{ in.}^2/\text{ft}$$

For main reinforcement, use No. 4 bars at 9-in. centers ($= 0.27$ in.²/ft).
    Deflection: We must check whether the slab satisfies the deflection criterion set out in Table 8.2.   The slab has one end discontinuous, and therefore requires

$$h \geq \frac{\ell_n}{24} = \frac{13.5 \times 12}{24} = 6.75 \text{ in.}$$

The slab must therefore be made thicker. *Use a 7-in. slab.* We could reduce the main reinforcement for this thicker slab, but the reinforcement is already light, and the revision is not warranted.
    Secondary reinforcement: The slab requires reinforcement to resist temperature and shrinkage stresses (Section 6.3), which is

$$0.001\,8 \times 12 \times 7 = 0.151 \text{ in.}^2/\text{ft}$$

For secondary reinforcement, use No. 4 bars at 15-in. centers ($= 0.16$ in.²/ft).
    Additional checks: The following criteria must also be satisfied:

1. The maximum spacing of the main reinforcement must not exceed (Section 6.4)

$$3 \times 7 = 21 \text{ in. or } 18 \text{ in.}   \text{O.K.}$$

2. The minimum reinforcement in either direction is

$$0.001\,8 \times 12 \times 7 = 0.151 \text{ in.}^2/\text{ft}   \text{O.K.}$$

3. The weight of a 7-in.-thick reinforced concrete slab is (Section 7.1)

$$7 \times 12 = 84 \text{ psf}$$

The allowance for dead load $D = 100$ psf, which allows 16 psf for finishes.   O.K.

    Reinforcement at other sections: The maximum positive bending moment (Fig. 7.4) is $+ w_u \ell^2_n/14$, and there are negative moments at the outer supports equal to $- w_u \ell_n^2/24$.   We could reduce the reinforcement for these smaller banding moments, but we are already using the smallest bars, and using different spacing in different parts of the sample slab causes construction problems, unless the wider spacing is twice the narrower spacing.   We will therefore *use No. 4 bars at 9-in. centers for positive moment* and we will *use No. 4 bars at 18-in. centers for negative moment at the exterior supports*.

Anchorage and bending up of reinforcement: The rules for anchorage are set out in Section 6.7.   The rules are complicated, but they are satisfied if we take half the positive reinforcement to the supports of each span, and bend the other half up to use as negative reinforcement.   The layout of the reinforcement is shown in Fig. 9.3(a), and the bars to be used are drawn separately in Fig. 9.3(b).

Half the positive steel is bent up to provide the small amount of reinforcement needed over the outer supports (bars Type C and D).   The other half is bent up to provide the negative reinforcement over the central support (bars Type A and B).

The negative reinforcement over the central supports is provided by alternate bars A and B, each type on 18-in. centers, giving an effective spacing of 9 in.   The positive reinforcement is provided by alternate bars A and C in the left-hand span, and by alternate bars B and D in the right-hand span.   The negative reinforcement for the left-hand support consists of bars C on 18-in. centers; that is also their effective spacing, because the bending moment over the outer supports is much lower.

Development length: There is insufficient space for a straight development length at the outer supports, and bars C and D are therefore bent at their ends; 90° hooks are more suitable because they do not interfere with the reinforcement of the supporting beams.

Half the positive reinforcement is taken to the face of the support, as shown in Fig. 9.3(a), and the other half is bent up.   The bars are of small diameter, and a glance at Table 6.1 shows that we have adequate development length without the need

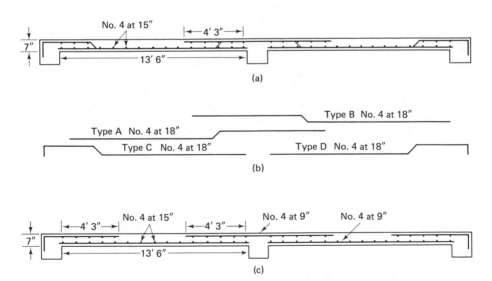

Figure 9.3 Arrangement of the reinforcement for the two-span continuous slab of Example 9.1.   (a) Section showing bar layout if reinforcement is bent up.   (b) The four types of bar used in Fig. 9.3(a).   (c) Section showing bar layout if only straight bars are used.

for a numerical check. Half the negative reinforcement over the central support is bent down, and the other half is taken a distance ℓ_a past the point of inflection.

The point of inflection (see Fig. 6.8) could be calculated from the bending moment diagram, but its exact location is not important, provided that we estimate it conservatively. One-fourth of the span from the support is a conservative estimate.

$$\tfrac{1}{4}\ell_n = 3 \text{ ft } 4\tfrac{1}{2} \text{ in.}$$

The additional length is

$$\ell_a \geqslant d = 7 - 0.75 - 0.25 = 6 \text{ in.}$$

$$\geqslant 12d_b = 12 \times 0.5 = 6 \text{ in.}$$

$$\geqslant \tfrac{1}{16}\ell_n = 0.844 \text{ ft } = 10\tfrac{1}{2} \text{ in.}$$

Therefore, the bars must extend 4 ft 3 in. beyond the face of the support.

Secondary reinforcement: The secondary reinforcement that resists temperature and shrinkage stresses does not require anchorage. It is very useful for tying the main reinforcing bars together, so that the reinforcement in both directions is transformed into a single cage supported off the formwork.

We have calculated the spacing of this reinforcement (No. 4 bars on 15-in. centers) on the basis that there is one layer of bars. When reinforcement occurs on both faces in the zone 4 ft 3 in. from the face of each support, the secondary reinforcement could, in theory, be halved. However, the saving in material probably does not compensate for the extra effort resulting from the change in the bar spacing in the case of this small structure.

Alternative arrangement of main reinforcement with straight bars: The same reinforcement can be provided by using separate straight bars for each of the three bending moments as shown in Fig. 9.3(c). This layout uses slightly more steel, and the negative reinforcement at the top of the slab is not automatically supported off the bottom bars. But the appreciable cost of bar bending is saved.

All the details are the same as for the bent-up reinforcement layout. We could cut off one-half or two-thirds of the straight bars when they are no longer required, but the extra complication is probably not warranted for this small structure and for No. 4 bars. Therefore, all the positive reinforcement is taken to the face of the support. All the negative reinforcement for all three supports is taken 4 ft 3 in. beyond the face of the support.

Example 9.3A

Determine the maximum permissible live load for the slab shown in Fig. 9.3. The dead load is 100 psf, the concrete is Grade 3.5, and the steel is Grade 60.

Data given: $D = 100$ psf, $\ell_n = 13.5$ ft, $b = 1$ ft $= 12$ in., $f'_c = 3,500$ psi, $f_y = 60,000$ psi, $h = 7$ in., and $\phi = 0.90$.

Reinforcement at point of maximum bending moment $=$ No. 4 bars on 9-in. centers. From Table C3,

$$A_s = 0.27 \text{ in.}^2/\text{ft}$$

Effective depth, from Eq. (6.1), is

$$d = 7 - 0.75 - 0.25 = 6 \text{ in.} = 0.5 \text{ ft}$$

The reinforcement ratio is

$$\rho = \frac{0.27}{12 \times 6} = 0.00375$$

From Table 9.1,

$$\frac{z}{d} = 0.96$$

From Eq. (9.8), the resistance moment is

$$M' = \phi f_y A_s z = 0.90 \times 60,000 \text{ psi} \times 0.27 \text{ in.}^2/\text{ft} \times 0.96 \times 0.5 \text{ ft}$$

$$= 7,000 \text{ lb} \cdot \text{in.}/\text{in.} = 7,000 \text{ lb} \cdot \text{ft/ft}$$

This equals the ultimate bending moment, which, from Fig. 7.4, is

$$M_u = \frac{-w_u \ell_n^2}{9} = 7,000 \text{ lb} \cdot \text{ft/ft}$$

Therefore, the ultimate load per unit area of slab is

$$w_u = \frac{7,000 \times 9}{13.5^2} = 345.6 \text{ psf}$$

The factored dead load is $1.4 \times 100 = 140$ psf, which leaves 205.6 psf for the factored live load. Since the load factor is 1.7, the service live load

$$L = \frac{205.6}{1.7} = 121 \text{ psf}$$

We designed the slab for a live load of 60 psf. This is now twice as high partly because we used a thicker slab to satisfy the deflection criterion, and partly because of the general increase upward to use round figures for the bar spacing and the slab thickness.

Example 9.3B

Determine the maximum permissible live load for the slab designed in Example 9.1B.

   Data given: $D = 5$ kPa, $\ell_n = 4$ m, $b = 1$ m, $f_c' = 25$ MPa, $f_y = 400$ MPa, $h = 175$ mm, and $\phi = 0.90$.

   Reinforcement at the point of maximum bending moment = No. 10 bars on 175-mm centers. From Table C4,

$$A_s = 571 \text{ mm}^2/\text{mm} = 571 \times 10^{-6} \text{ m}^2/\text{m}$$

Effective depth, from Eq. (6.1),

$$d = 175 - 20 - 5 = 150 \text{ mm}$$

The reinforcement ratio is

$$\rho = \frac{571}{1\,000 \times 150} = 0.0038$$

From Table 9.2,

$$\frac{z}{d} = 0.96$$

From Eq. (9.8), the resistance moment is

$$M' = \phi f_y A_s z = 0.90 \times 400 \times 571 \times 10^{-6} \times 0.96 \times 0.150$$

$$= 29.60 \times 10^{-3} \text{ MN} \cdot \text{m/m} = 29.60 \text{ kN} \cdot \text{m/m}$$

This equals the ultimate bending moment, which, from Fig. 7.4, is

$$M_u = \frac{-w_u \ell_n^2}{9} = 29.60 \text{ kN} \cdot \text{m/m}$$

Therefore, the ultimate load per unit area of slab

$$w_u = \frac{29.60 \times 9}{4.000^2} = 16.65 \text{ kPa}$$

The factored dead load is $1.4 \times 5 = 7$ kPa, which leaves 9.65 kPa for the factored live load. Since the load factor is 1.7, the service live load is

$$L = \frac{9.65}{1.7} = 5.68 \text{ kPa}$$

We designed the slab for a live load of 3 kPa. It is now 90% higher partly because we used a thicker slab to satisfy the deflection criterion, and partly because of the general increase upward to use round figures for the bar spacing and the slab thickness.

Example 9.4A

Design a one-way simply supported reinforced concrete slab over a span of 13 ft 6 in. to carry a live load of 60 psf using Grade 3.5 concrete and Grade 60 steel.

Data given: $L = 60$ psf, $\ell = 13.5$ ft, $b = 1$ ft, $f_c' = 3,500$ psi, $f_y = 60,000$ psi, and $\phi = 0.90$.

It is evident from Tables 8.1 and 8.2 that the deflection criterion is more significant for a simply supported slab than for a continuous slab. From Table 8.2, the minimum overall depth is

$$h = \frac{\ell}{20} = \frac{13.5 \times 12}{20} = 8.1 \text{ in.}$$

Use an 8½ in. slab.

The weight of an 8½-in.-thick slab is $8.5 \times 12 = 102$ psf (Section 7.1). Allowing for finishes, we will make $D = 110$ psf. The ultimate load per unit area of slab

$$w_u = 1.4 \times 110 + 1.7 \times 60 = 256 \text{ psf}$$

The maximum bending moment (Table 7.3) is

$$M_u = \frac{256 \times 13.5^2}{8} = 5,832 \text{ lb} \cdot \text{ft/ft} = 69,984 \text{ lb} \cdot \text{in./ft}$$

Let us assume conservatively that $z/d = 0.95$. The effective depth is

$$d = 8.5 - 0.75 - 0.25 \text{ (for a No. 4 bar)} = 7.5 \text{ in.}^z$$

From Eq. (9.8) the resistance moment of the slab is

$$M' = \phi f_y A_s z$$

which gives

$$A_s = \frac{69,984}{0.90 \times 60,000 \times 0.95 \times 7.5} = 0.182 \text{ in.}^2/\text{ft}$$

Use No. 4 bars at 12-in. centers ($= 0.20$ in.²/ft).

The temperature and shrinkage reinforcement is

$$0.0018 \times 12 \times 7 = 0.151 \text{ in.}^2/\text{ft}$$

Use No. 4 bars at 15-in. centers ($= 0.16$ in.²/ft).

We have checked the weight and the deflection. The maximum permissible primary bar spacing is 18 in. (We have 12 in.) O.K.

$$\rho = \frac{0.20}{12 \times 7} = 0.0024 \quad \text{and} \quad \frac{z}{d} = 0.97 \quad \text{(we assumed 0.95)} \quad \text{O.K.}$$

From Table 6.1, the development length is clearly adequate for No. 4 bars, but there is insufficient room for the additional length ℓ_a required beyond the point of zero bending moment, which occurs at the support. Hooks are required. These can be turned down into the beam, but may interfere with beam reinforcement. It is better to turn them up into the slab. 180° hooks require a depth of (Fig. 6.6) $8d_b =$ 4 in. for No. 4 bars; 90° hooks require $16d_b = 8$ in., which is not available in an 8½-in. slab.

The layout of the reinforcement is shown in Fig. 9.4.

Example 9.4B

Design a one-way simply supported reinforced concrete slab over a span of 4 m, to carry a live load of 3 kPa, using Grade 25 concrete and Grade 400 steel.

Data given: $L = 3$ kPa, $\ell = 4$ m, $b = 1$ m, $f_c' = 25$ MPa, $f_y = 400$ MPa, and $\phi = 0.90$.

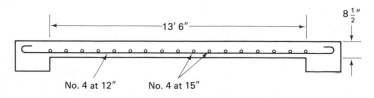

Figure 9.4 Layout of the reinforcement for the slab of Example 9.8.

It is evident from Tables 8.1 and 8.2 that the definition criterion is more significant for a simply supported slab than for a continuous slab. From Table 8.2, the minimum overall depth is

$$h = \frac{\ell}{20} = \frac{4000}{20} = 200 \text{ mm}$$

Use a 200-mm slab.

The weight of a 200-mm thick slab is $0.200 \times 24 = 4.8$ kPa. Allowing for finishes, we will make $D = 5.2$ kPa. The ultimate load per unit area of slab is

$$w_u = 1.4 \times 5.2 + 1.7 \times 3 = 12.38 \text{ kPa}$$

The maximum bending moment (Table 7.3) is

$$M_u = \frac{12.38 \times 4^2}{8} = 24.76 \text{ kN} \cdot \text{m/m} = 24.76 \times 10^{-3} \text{ MN} \cdot \text{m/m}$$

Let us assume conservatively that $z/d = 0.95$. The effective depth is

$$d = 200 - 20 - 5 \text{ (for a No. 10 bar)} = 175 \text{ mm} = 0.175 \text{ m}$$

From Eq. (9.8) the resistance moment of the slab is

$$M' = \phi f_y A_s z$$

which gives

$$A_s = \frac{24.76 \times 10^{-3}}{0.90 \times 400 \times 0.95 \times 0.175} = 414 \times 10^{-6} \text{ m}^2/\text{m} = 414 \text{ mm}^2/\text{m}$$

Use No. 10 bars at 200-mm centers ($= 500$ mm²/m).

The temperature and shrinkage reinforcement is

$$0.0018 \times 1\,000 \times 200 = 360 \text{ mm}^2/\text{m}$$

Use No. 10 bars at 250-mm centers ($= 400$ mm²/m).

We have checked the weight and the deflection. The maximum permissible primary bar spacing is 450 mm. (We have 200 mm.) O.K.

$$\rho = \frac{500}{1\,000 \times 175} = 0.0029 \quad \text{and} \quad \frac{z}{d} = 0.97 \quad \text{(we assumed 0.95)} \qquad \text{O.K.}$$

9.3 PRIMARY TENSION AND COMPRESSION FAILURE

The failure of a reinforced concrete beam or slab may be initiated by the yielding of the steel, followed by the rise of the neutral axis, which eventually produces crushing of the concrete. This is called a *primary tension failure* (Fig. 9.5). In a lightly reinforced beam the depth of the neutral axis at failure is quite small. For example, for a yield stress $f_y = 60,000$ psi, a concrete crushing strength $f'_c = 4\,000$ psi ($\beta_1 = 0.85$, see Fig. 8.5), and a reinforcement

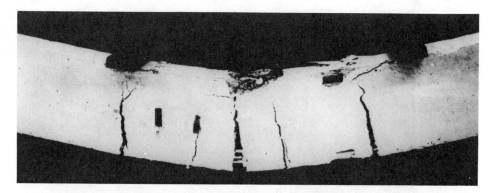

Figure 9.5   Primary tension failure of reinforced concrete beam.   Note the width of the cracks at failure and the small depth of the neutral axis.

ratio $\rho = 0.006$, the neutral-axis depth ratio at failure, from Eq. (9.4), is

$$\frac{c}{d} = \frac{60{,}000}{0.85 \times 400} \times \frac{0.006}{0.85} = 0.125$$

This means that the depth of the neutral axis is only 12½% of the effective depth.   Failure is gradual as the steel strain increases and the neutral axis rises.

The failure of a reinforced beam or slab may also be initiated by crushing of the concrete before the steel yields.   This is called a *primary compression failure*, and it occurs when the neutral axis is still at approximately half-depth (Fig. 9.6).   This type of failure is sudden, and there is little warning of impending failure.   The *ACI Code* does not therefore permit a design that would result in primary compression failure.

A primary compression failure can be prevented by setting an upper limit to the reinforcement ratio.   To determine this limit we examine the *balanced failure condition*, at which the concrete crushes and the steel begins to yield at precisely the same bending moment (i.e., a primary compression failure and a primary tension failure occur simultaneously).

A balanced failure is reached when the maximum concrete strain ε_c' and the steel strain ε_y occur at the same bending moment (Fig. 9.7). This means that

$$\frac{c}{d - c} = \frac{\varepsilon_c'}{\varepsilon_y} \tag{9.9}$$

which gives the balanced neutral-axis depth

$$\frac{c_b}{d} = \frac{\varepsilon_c'}{\varepsilon_c' + \varepsilon_y} \tag{9.10}$$

Figure 9.6 Primary compression failure of reinforced concrete beam. Note the comparative lack of tension cracks and the appreciable depth of the compression zone in which the concrete is crushed.

Substituting from Eq. (9.4) yields

$$\frac{c_b}{d} = \frac{f_y}{0.85f'_c}\frac{\rho_b}{\beta_1} = \frac{\varepsilon'_c}{\varepsilon'_c + \varepsilon_y} \tag{9.11}$$

Thus the limiting reinforcement ratio for a balanced failure condition is

$$\rho_b = \frac{\varepsilon'_c}{\varepsilon'_c + \varepsilon_y}\, 0.85\beta_1 \frac{f'_c}{f_y} \tag{9.12}$$

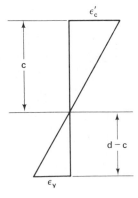

Figure 9.7 Depth of neutral axis at the balanced failure condition. $\varepsilon'_c =$ ultimate concrete strain $= 0.003$ for all values of f'_c; $\varepsilon_y =$ yield strain of steel $= f_y/E_s$, where $E_s = 29{,}000{,}000$ psi (200 000 MPa).

TABLE 9.3 Balanced Neutral Axis Depth Ratio c_b/d, and Balanced Reinforcement Ratio ρ_b for Various Grades of Steel (Yield Stress f_y) and Concrete (Specified Compressive Strength f_c')

(a) In Pounds per Square Inch (psi)				(b) In Megapascals (MPa)			
f_y	f_c'	$\dfrac{c_b}{d}$	ρ_b	f_y	f_c'	$\dfrac{c_b}{d}$	ρ_b
40,000	3000	0.685	0.0371	300	20	0.666	0.0320
	4000	0.685	0.0495		25	0.666	0.0401
	5000	0.685	0.0560		30	0.666	0.0481
	6000	0.685	0.0603		35	0.666	0.0535
60,000	3000	0.592	0.0214	400	20	0.600	0.0216
	4000	0.592	0.0285		25	0.600	0.0271
	5000	0.592	0.0323		30	0.600	0.0325
	6000	0.592	0.0347		35	0.600	0.0361

The *ACI Code* specifies $\varepsilon_c' = 0.003$ for all values of f_c' (Fig. 8.4). The yield strain

$$\varepsilon_y = \frac{f_y}{E_s}$$

and the modulus of elasticity $E_s = 29{,}000{,}000$ psi ($200\ 000$ MPa) for all values of f_y. The *ACI Code* also specifies the value of β_1 (Fig. 8.5). We can thus compute the limiting reinforcement ratio for the various grades of steel and concrete; see Table 9.3.

9.4 LIMITS ON REINFORCEMENT RATIOS

To ensure that primary compression failure does not occur, the *ACI Code* (Section 10.3.3) limits the neutral-axis depth ratio to 75% of the balanced neutral-axis depth ratio; that is,

$$c_{\max} \leq 0.75 c_b$$

Consequently, the maximum permissible value, from Eq. (9.10), is

$$\frac{c_{\max}}{d} = 0.75 \frac{\varepsilon_c'}{\varepsilon_c' + \varepsilon_y} \tag{9.13}$$

and the maximum permissible reinforcement ratio [from Eq. (9.12)] is

$$\rho_{\max} = 0.75 \frac{\varepsilon_c'}{\varepsilon_c' + \varepsilon_y} 0.85 \beta_1 \frac{f_c'}{f_y} \tag{9.14}$$

TABLE 9.4 Maximum Permissible Neutral-Axis Depth c_{max} and Maximum Permissible Reinforcement Ratio ρ_{max} for Various Grades of Steel and Concrete at the Ultimate Load

(a) In Pounds per Square Inch (psi)				(b) In Megapascals (MPa)			
f_y	f'_c	$\dfrac{c_{max}}{d}$	ρ_{max}	f_y	f'_c	$\dfrac{c_{max}}{d}$	ρ_{max}
40,000	3000	0.514	0.0278	300	20	0.500	0.0240
	4000	0.514	0.0371		25	0.500	0.0301
	5000	0.514	0.0420		30	0.500	0.0361
	6000	0.514	0.0452		35	0.500	0.0401
60,000	3000	0.444	0.0161	400	20	0.450	0.0162
	4000	0.444	0.0214		25	0.450	0.0203
	5000	0.444	0.0242		30	0.450	0.0244
	6000	0.444	0.0260		35	0.450	0.0271

These values are listed in Table 9.4.   Any reinforcement ratio greater than ρ_{max} is not admissible.   Hence the corresponding values of z/d have been omitted from Tables 9.1 and 9.2.

The maximum permissible ultimate resistance moment for a beam or slab of width b and effective depth d is, therefore, from Eq. (9.8),

$$M'_{max} = \phi f_y \rho_{max} \frac{z}{d} bd^2 = k_m bd^2 \tag{9.15}$$

where

$$k_m = \phi f_y \rho_{max} \frac{z}{d} \tag{9.16}$$

The limiting values for the reinforcement ratio ρ, the moment arm ratio, and the values of the design constant k_m are listed in Table 9.5.

The brittleness of concrete is one of the main causes of the occasional failure of reinforced concrete structures in earthquakes.   It is, therefore, desirable that the reinforcement ratio for structures in seismic zones (see Fig. 1.3) be further limited, say to $0.50\rho_b$, instead of $0.75\rho_b$.   For small buildings this limitation presents no problems.

Example 9.5

Redesign the reinforced concrete slab of Examples 9.1A. and 9.2 for minimum depth, using the slightly stronger Grade 4 concrete.

Data given: $D = 100$ psf, $L = 60$ psf, $\ell_n = 13.5$ ft, $b = 1$ ft, $f'_c = 4,000$ psi, $f_y = 60,000$ psi, $k_m = 881$ psi, and $\rho_{max} = 0.0214$.

Since the slab will be much thinner, the dead load D is too high.   However,

TABLE 9.5 Design Constants for the Maximum Permissible Ultimate
Resistance Moment

(a) In Pounds per Square Inch (psi)					(b) In Megapascals (MPa)				
f_y (psi)	f'_c (psi)	ρ_{max} (ratio)	$\dfrac{z}{d}$ (ratio)	k_m (psi)	f_y (MPa)	f'_c (MPa)	ρ_{max} (ratio)	$\dfrac{z}{d}$ (ratio)	k_m (MPa)
40,000	3000	0.0278	0.778	735	300	20	0.0240	0.787	5.11
	4000	0.0371	0.778	981		25	0.0301	0.787	6.40
	5000	0.0420	0.798	1,140		30	0.0361	0.787	7.68
	6000	0.0452	0.819	1,259		35	0.0401	0.797	8.65
60,000	3000	0.0161	0.807	663	400	20	0.0162	0.809	4.73
	4000	0.0214	0.807	881		25	0.0203	0.809	5.92
	5000	0.0242	0.826	1,019		30	0.0244	0.809	7.10
	6000	0.0262	0.844	1,119		35	0.0271	0.818	7.98

we will leave it as it is, because this example is an academic exercise to show that the design of *slabs* at the limiting moment M'_{max} is impracticable.

$$w_u = 1.4 \times 100 + 1.7 \times 60 = 242 \text{ psf}$$

$$M_u = \frac{-w_u \ell_n^2}{9} = \frac{-242 \times 13.5^2}{9} = -4,900 \text{ lb} \cdot \text{ft/ft}$$

$$= -4,900 \text{ lb} \cdot \text{in./in.}$$

From Eq. (9.15),

$$M'_{max} = M_u = k_m b d^2$$

$$4,900 = 881 \times 1 \times d^2$$

$$d = 2.36 \text{ in.}$$

$$A_s = \rho_{max} b d = 0.0214 \times 12 \times 2.36 = 0.606 \text{ in.}^2/\text{ft}$$

Use No. 5 bars at 6-in. mm centers.
  The overall depth is

$$h = 2.36 + 0.31 + 0.75 = 3.42 \text{ in.}, \quad say \ 3\tfrac{1}{2} \text{ in.}$$

The minimum overall depth required by the approximate method set out in Table 8.2 is

$$\frac{\ell_n}{24} = \frac{13.5 \times 12}{24} = 6.75 \text{ in.}$$

so that the slab is not thick enough. Please note that Table 8.2 is applicable only to slabs not connected to partitions or other construction likely to be damaged by large deflections.
  A more accurate result can be obtained by calculating the moment of inertia of the cracked reinforced concrete section, and using it in Eq. (8.4) to calculate the

deflection. This requires a lengthy calculation, which is explained in more advanced textbooks (e.g., Ref. 7.5). However, in this example the slab still fails the deflection criterion when an accurate calculation is made. It is generally not possible to design a reinforced concrete slab reinforced with ρ_{max}, without failing the deflection criterion.

The high percentages of reinforcement associated with maximum admissible resistance moments are, however, useful in the design of beams, where they produce the shallowest beams attainable without compression reinforcement (see Sections 9.6 and 10.4).

The simply supported slab of Example 9.4A. is even less suitable for a high reinforcement ratio, because of the high deflection resulting from the simple supports.

The bending moment calculated in Example 9.4,

$$M_u = 69,984 \text{ lb} \cdot \text{in./ft}$$

produces an effective depth, from Eq. (9.15), of

$$d = \sqrt{\frac{69,984}{881 \times 12}} = 2.57 \text{ in.}$$

whereas the deflection criterion requires that

$$h = \frac{\ell}{20} = 8.1 \text{ in.}$$

9.5 CHOICE OF REINFORCEMENT RATIO

The deflection criterion generally determines the thickness of slabs. The coefficients of Table 8.2 give a simple answer for slabs not connected to partitions. Thinner slabs can be attained by determining the deflection accurately from Eq. (8.4), but the calculations are lengthy.

Having determined the depth, the area of reinforcement, from Eq. (9.8), is

$$A_s = \frac{M_u}{\phi f_y z} \tag{9.17}$$

Beams can be designed at the upper limit of reinforcement ρ_{max}, but if depth is no problem, a more economical design is often obtained by using a rule given in the 1963 *ACI Code* (Table 9.6):

$$\rho = \frac{0.18 f_c'}{f_y} \tag{9.18}$$

Thus, from Eq. (9.8) the product

$$bd^2 = \frac{M_u}{\phi f_y \rho} \frac{d}{z} \tag{9.19}$$

If $\rho = 0.18 f_c'/f_y$ is used, z/d (from Table 9.6) has a constant value 0.89. The

TABLE 9.6   Values of $\rho = 0.18 f'_c/f_y$ and of the Corresponding z/d

(a) In Pounds per Square Inch (psi)				(b) In Megapascals (MPa)			
f_y	f'_c	$\rho = 0.18\dfrac{f'_c}{f_y}$	$\dfrac{z}{d}$	f_y	f'_c	$\rho = 0.18\,f'_c/f_y$	z/d
40,000	3000	0.014	0.89	300	20	0.012	0.89
	4000	0.018	0.89		25	0.015	0.89
	5000	0.023	0.89		30	0.018	0.89
	6000	0.027	0.89		35	0.021	0.89
60,000	3000	0.009	0.89	400	20	0.009	0.89
	4000	0.012	0.89		25	0.011	0.89
	5000	0.015	0.89		30	0.014	0.89
	6000	0.018	0.89		35	0.016	0.89

ratio of width to depth is at the designer's choice: a ratio d/b from 1.5 to 2.5 is suitable.

The area of reinforcement is

$$A_s = \rho b d \tag{9.20}$$

9.6 DESIGN OF RECTANGULAR BEAMS

Example 9.6A

Design a simply supported rectangular beam carrying a dead load of 1000 lb/ft and a live load of 500 lb/ft simply supported over a span of 30 ft, using Grade 4 concrete and Grade 60 steel.

Data given: $D = 1000$ lb/ft, $L = 500$ lb/ft, $\ell = 30$ ft, $f'_c = 4000$ psi, $f_y = 60,000$ psi, and $\phi = 0.90$.

The ultimate load is

$$w_u = 1.4 \times 1000 + 1.7 \times 500 = 2250 \text{ lb/ft run of beam}$$

The ultimate bending moment for the simply supported beam is

$$M_u = \frac{2250 \times 30^2}{8} = 253,125 \text{ lb} \cdot \text{ft}$$

$$= 3,038,000 \text{ lb} \cdot \text{in.}$$

We will first design the beam using Eq. (9.18):

$$\rho = \frac{0.18 f'_e}{f_y} = 0.012$$

which gives (Table 9.6)

$$\frac{z}{d} = 0.89$$

From Eq. (9.19) the product is

$$bd^2 = \frac{3{,}038{,}000}{0.90 \times 60{,}000 \times 0.012 \times 0.89} = 5267 \text{ in.}^3$$

If $d = 2b$,

$$b^3 = 13.17 \text{ in.}^3 \quad \text{and} \quad b = 10.96 \text{ in.}$$

The width needs to be a round figure, whereas the effective depth need not be. *Use* $b = 12$ *in.*

$$d^2 = \frac{5267}{12} = 438.9 \text{ in.}^2 \quad \text{and} \quad d = 20.95 \text{ in.}$$

$$A_s = 0.012 \times 12 \times 20.95 = 3.017 \text{ in.}^2$$

Use 4 No. 8 bars ($= 3.16$ *in.*^{2}).

The beam requires nominal shear reinforcement, whose design is discussed in Section 11.5. We must add an allowance for a stirrup diameter of 3/8 in. in determining the overall depth.

The overall depth

$$h = d + \tfrac{1}{2} \text{ bar diameter (1 in.)} + 1 \text{ stirrup diameter (}\tfrac{3}{8}\text{ in.)} + \text{cover}$$

$$= 20.95 + 0.50 + 0.38 + 1.50 = 23.33 \text{ in.}$$

Use a 12 in. by 24 in. beam.

The deflection criterion (Table 8.2) requires a minimum depth

$$h = \frac{\ell}{20} = \frac{30 \times 12}{20} = 18 \text{ in. (24 in.)} \quad \text{O.K.}$$

The minimum positive reinforcement ratio (Section 6.3) required is

$$\rho = \frac{200}{f_y} = 0.003\,3 \ (0.011\,2) \quad \text{O.K.}$$

For the three clear spaces between the four bars we have

$$b - 4 \text{ No. 8 bar diameters} - 2 \text{ stirrup (No. 3) diameters} - 2 \times \text{cover}$$

$$= 12 - 4 \times 1.0 - 2 \times 0.38 - 2 \times 1.5 = 4.25 \text{ in.}$$

This gives a 1.42-in. clear space between bars. We require a minimum of 1 in. (irrespective of bar size) and also at least the bar diameter ($d_b = 1$ in.). O.K.

From Table 6.1, development length is clearly adequate for No. 8 bars, but there is insufficient room for the additional length ℓ_a required beyond the point of zero bending moment, which occurs at the support. Hooks are required; either 180° hooks, 8 in. deep, or 90° hooks, $16d_b = 16$ in. deep (Fig. 6.6) can be used.

In a simply supported beam the maximum positive bending moment occurs at midspan and reduces to zero at the supports. The negative bending moment is theoretically zero throughout, but if the ends of the beam are restrained because, say, a partition rests on them, there is a small negative bending moment at the supports. It is therefore appropriate, since the full reinforcement is not required at the bottom at the supports, to turn two bars up at the ends to take care of any negative bending moment.

The layout of the reinforcement is shown in Fig. 9.8.   The shear reinforcement is designed in Section 11.5.

Example 9.6B

Design a simply supported rectangular beam carrying a dead load of 12 kN/m and a live load of 7 kN/m simply supported over a span of 10 m, using Grade 25 concrete and Grade 400 steel.

Data given: $D = 12$ kN/m, $L = 7$ kN/m, $\ell = 10$ m, $f_c' = 25$ MPa, $f_y = 400$ MPa, and $\phi = 0.90$.

The ultimate load is

$$w_u = 1.4 \times 12 + 1.7 \times 7 = 28.7 \text{ kN per meter run of beam}$$

The ultimate bending moment for the simply supported beam is

$$M_u = \frac{28.7 \times 10^2}{8} = 358.75 \text{ kN} \cdot \text{m}$$

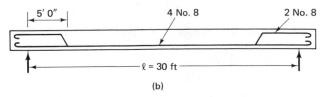

(a)

(b)

Figure 9.8   Reinforcement details for the beam of Example 9.6.   (a) Cross section. (b) Elevation (stirrups not shown).

We will first design the beam using Eq. (9.18):

$$\rho = \frac{0.18f'_c}{f_y} = 0.011$$

which gives (Table 9.6)

$$\frac{z}{d} = 0.89$$

From Eq. (9.19) the product is

$$bd^2 = \frac{358.75 \times 10^{-3}}{0.90 \times 400 \times 0.011 \times 0.89} = 101.79 \times 10^{-3}\, \text{m}^3$$

If $d = 2b$,

$$b^3 = 25.45 \times 10^{-3}\, \text{m}^3 \quad \text{and} \quad b = 0.294\, \text{m}$$

The width needs to be a round figure, whereas the effective depth need not be. *Use b = 300 mm.*

$$d^2 = \frac{101.79 \times 10^{-3}}{0.3} = 339.3 \times 10^{-3}\, \text{m}^2 \quad \text{and} \quad d = 0.582\, \text{m} = 582\, \text{mm}$$

$$A_s = 0.011 \times 300 \times 582 = 1\,921\, \text{mm}^2$$

Use 4 No. 25 bars ($= 2\,000$ mm²).

The beam requires nominal shear reinforcement, whose design is discussed in Section 11.5. We must add an allowance for a stirrup diameter of 10 mm in determining the overall depth. The overall depth is

$$h = d + \tfrac{1}{2}\, \text{bar diameter (25 mm)}$$

$$+ \; 1\, \text{stirrup diameter (10 mm)} + \text{cover}$$

$$= 582 + 13 + 10 + 40$$

$$= 645\, \text{mm}$$

Use a 300-mm by 650-mm beam.

The deflection criterion (Table 8.2) requires a minimum depth

$$h = \frac{\ell}{20} = \frac{10\,000}{20} = 500\, \text{mm} \quad (650\, \text{mm}) \quad \text{O.K.}$$

The minimum positive reinforcement ratio (Section 6.3) required is

$$\rho = \frac{1.4}{f_y} = 0.003\,5 \quad (0.011) \quad \text{O.K.}$$

For the three clear spaces between the four bars we have

$$b - 4\, \text{No. 25 bar diameters} - 2\, \text{stirrup (No. 10) diameters} - 2 \times \text{cover}$$

$$= 300 - 4 \times 25 - 2 \times 10 - 2 \times 40 = 100\, \text{mm}$$

This gives a 33-mm clear space between bars. We require a minimum of 25 mm (irrespective of bar size) and also at least the bar diameter ($d_b = 25$ mm). O.K.

Example 9.7

Design the beam of Example 9.6 for minimum depth, but without the use of compression reinforcement (see Section 10.4).

   Data given: $M_u = 3{,}038{,}000$ lb $\cdot$ in., $\ell = 30$ ft, $f'_c = 4000$ psi, $f_y = 60{,}000$ psi, $\phi = 0.90$, $k_m = 881$ psi, $\rho_{max} = 0.0214$, and $z/d = 0.807$.

   The deflection criterion

$$h = \frac{\ell}{20} = \frac{30 \times 12}{20} = 18 \text{ in.}$$

limits the depth, unless more accurate deflection calculations are made.   Assuming that $h = 18$ in., one layer of No. 10 bars, and ⅜-in. stirrups,

$$d = 18 - 1.5 - 0.38 - \tfrac{1}{2} \times 1.27 = 15.5 \text{ in.}$$

From Eq. (9.15),

$$M'_{max} = M_u = k_m bd^2$$

which gives the width

$$b = \frac{3{,}038{,}000}{881 \times 15.5^2} = 14.35 \text{ in.}$$

Use b = 12 in.
This gives a correct effective depth of

$$d = \sqrt{\frac{3{,}038{,}000}{881 \times 12}} = 16.95 \text{ in.}$$

From $\rho = 0.0214$, the area of reinforcement required is

$$A_s = 0.0214 \times 12 \times 16.95 = 4.353 \text{ in.}^2$$

Use 4 No. 10 bars (= 5.08 in.²) or 3 No. 11 bars (= 4.68 in.²).
Minimum width required for 4 No. 10 bars is

$b = 2 \times 1.5 \text{ (cover)} + 2 \times 0.38 \text{ (ties)}$

$$+ 4 \times 1.27 \text{ bars} + 3 \times 1.27 \text{ spaces} = 12.64 \text{ in.}$$

which is not available.   Therefore, it is necessary to use 3 No. 11 bars, which require

$$b = 3 + 0.75 + 3 \times 1.56 + 2 \times 1.56 = 11.55 \text{ in.}$$

Use 3 No. 11 bars.
The overall depth

$$h = 16.95 + \tfrac{1}{2} \times 1.56 + 0.38 + 1.5 = 19.67 \text{ in.}$$

Use a 12 in. by 20 in. beam.
   Neglecting the rounding of dimensions in both cases, this beam requires 4.352 in.² of steel (versus 3.017 in.² in Example 9.6), and an effective depth of 16.95 in. (versus 20.95 in.) for the same width.   That is 44% more steel for only 24% less concrete.

Example 9.8A

Determine the maximum permissible live load for the beam shown in Fig. 9.8.   The dead load is 1000 lb/ft, the concrete is Grade 4, the steel is Grade 60, and the beam is simply supported over a span of 30 ft.

   Data given: D = 1000 lb/ft, ℓ = 30 ft, b = 12 in., h = 24 in., f'_c = 4000 psi, f_y = 60,000 psi, ϕ = 0.90, A_s = 4 No. 8 bars.

$$A_s = 3.16 \text{ in.}^2$$

$$d = 24 - 1.5 - 0.38 - \tfrac{1}{2} \times 1.0 = 21.62 \text{ in.}$$

$$\rho = \frac{3.16}{12 \times 21.62} = 0.0122$$

From Table 9.1,

$$\frac{z}{d} = 0.89 \quad \text{and} \quad z = 0.89 \times 21.62 = 19.24 \text{ in.}$$

From Eq. (9.8),

$$M' = \phi f_y A_s z = 0.90 \times 60,000 \times 3.16 \times 19.24 = 3,283,000 \text{ lb} \cdot \text{in.}$$

$$= 273,600 \text{ lb} \cdot \text{ft}$$

$$\frac{w_u \ell^2}{8} = 273,600 \text{ lb} \cdot \text{ft}$$

The ultimate load per meter run is

$$w_u = \frac{273,600 \times 8}{30^2} = 2432 \text{ lb/ft}$$

The factored dead load is $1.4 \times 1000 = 1400$ lb/ft, leaving 1032 lb/ft for the factored live load.

$$L = \frac{1032}{1.7} = 607 \text{ lb/ft}$$

The beam was originally designed for a live load of 500 lb/ft.

Example 9.8B

Determine the maximum permissible live load for the beam shown in Fig. 9.8.   The dead load is 12 kN/m, the concrete is Grade 25, the steel is Grade 400, and the beam is simply supported over a span of 10 m.

   Data given: D = 12 kN/m, ℓ = 10 m, b = 300 mm, h = 650 mm, f'_c = 25 MPa, f_y = 400 MPa, ϕ = 0.90, A_s = 4 No. 25 bars.

$$A_s = 2\ 000 \text{ mm}^2$$

$$d = 650 - 40 - 10 - \tfrac{1}{2} \times 25 = 587 \text{ mm}$$

$$\rho = \frac{2\ 000}{300 \times 587} = 0.0114$$

From Table 9.2,

$$\frac{z}{d} = 0.89 \quad \text{and} \quad z = 0.89 \times 587 = 522 \text{ mm}$$

From Eq. (9.8),

$$M' = \phi f_y A_s z = 0.90 \times 400 \times 2\,000 \times 10^{-6} \times 0.522 = 375.8 \times 10^{-3} \text{ MN} \cdot \text{m}$$

$$\frac{w_u \ell^2}{8} = 375.8 \text{ kN} \cdot \text{m}$$

The ultimate load per meter run is

$$w_u = \frac{375.8 \times 8}{10^2} = 30.07 \text{ kN/m}$$

The factored deal load is $1.4 \times 12 = 16.8$ kN/m, leaving 13.27 kN/m for the factored live load.

$$L = \frac{13.27}{1.7} = 7.8 \text{ kN/m}$$

The beam was originally designed for a live load of 7 kN/m.

Example 9.9A

Design a rectangular beam continous over three equal spans of 30 ft, carrying a dead load of 1000 lb/ft and live load of 500 lb/ft, using Grade 4 concrete and Grade 60 steel.

Data given: $D = 1000$ lb/ft, $L = 500$ lb/ft, $\ell = 30$ ft (continuous over three spans), $f_c' = 4000$ psi, $f_y = 60,000$ psi, and $\phi = 0.90$.

Bending moments: Using the theory of continuous beams, the bending moment coefficients, from Table 7.1, are those in Fig. 9.9. The ultimate loads are

$$w_{uD} = 1.4 \times 1000 = 1400 \text{ lb/ft}$$

$$w_{uL} = 1.7 \times 500 = 850 \text{ lb/ft}$$

From Fig. 9.9, the ultimate bending moments are:

	Due to Dead Load	Due to Live Load	Total M_u (lb·ft)
Maximum negative	126,000	89,500	215,500
Positive, outer span	100,800	77,300	178,100
Positive, center span	31,500	57,400	88,900

Dimensions of section: We will first determine the effective depth for the maximum bending moment, using Eq. (9.18):

$$\rho = \frac{0.18 f_c'}{f_y} = 0.012$$

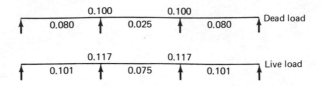

Figure 9.9 Bending moment coefficients for Example 9.9. M_u = coefficient $\times$ $w_u \ell^2$.

The separate coefficients for the dead and the live load are obtained in accordance with the *ACI Code* (see Section 7.3). The dead load must of necessity cover all spans ("if there is no dead load, there is no structure"). The *ACI Code* allows a limitation on the number of alternative loading conditions for the live load: (1) alternate spans loaded, which gives the highest positive moments, and (2) adjacent spans loaded, which gives the highest negative moments. The live-load bending moment coefficients are the highest obtained from this rule.

which (from Table 9.6), gives $z/d = 0.89$. From Eq. (9.19), the product

$$bd^2 = \frac{M_u}{\phi f_y \rho} \frac{d}{z}$$

$$= \frac{215,500 \times 12}{0.90 \times 60,000 \times 0.012 \times 0.89} = 4484 \text{ in.}^3$$

For $b = 12$ in., $d = 9.33$ in.

Assuming that the largest bar is No. 9, and that No. 3 stirrups are needed, the overall depth is

$$h = 19.33 + \tfrac{1}{2} \times 1.13 + 0.38 + 1.5 = 21.77 \text{ in.}$$

Use a 12 in by 22 in. beam.
The moment arm is

$$z = 0.89 \times 19.33 = 17.20 \text{ in.}$$

Reinforcement: The area of reinforcement at each section is

$$A_s = \frac{M_u}{\phi f_y z} = \frac{M_u \times 12}{0.90 \times 60,000 \times 17.20}$$

		Reinforcement
Negative	2.784 in.²	Use 2 No. 9 + 2 No. 6 = 2.88 in.²
Positive, outer span	2.301 in.²	Use 2 No. 9 + 2 No. 5 = 2.62 in.²
Positive, inner span	1.149 in.²	Use 4 No. 6 = 1.76 in.²

Detailing: The 2 No. 9 bars serve as negative reinforcement and as positive reinforcement in the outer span. The 2 No. 6 bars serve as negative reinforcement and as positive reinforcement in the center span. The center span has 2 No. 6 bars coming from the left and another 2 coming from the right, making 4 bars. This is

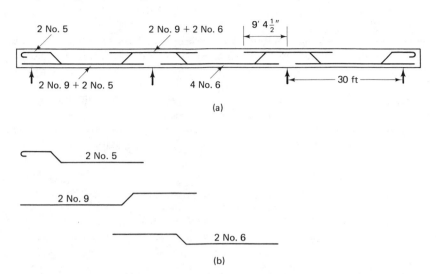

Figure 9.10   Reinforcement details for the beam of Example 9.9.   (a) Elevation (stirrups not shown).   (b) Types of main reinforcement used.

more than needed, but the bars are available because they are needed over the neighboring supports.   The 2 No. 5 bars which provide the additional positive reinforcement in the outer span, are bent up over the outer supports, where a small negative moment should be allowed for, although theoretically there is none.   The arrangement is shown in Fig. 9.10.

The critical combinations for *width* are 2 No. 9 + 2 No. 6 bars, which require $b = 2 \times 1.5$ (cover) + 2×0.38 (ties) + 2×1.13 (bars) + 2×0.75 (bars) + 2×1.13 (spaces between No. 9 and No. 6) + 1×0.75 (space between 2 No. 6) = 10.52 in.   (12 in.)   O.K.

Deflection criterion: From Table 8.2, for the outer span, the minimum depth is

$$h = \frac{\ell}{24} = \frac{30 \times 12}{24} = 15 \text{ in. (22 in.)   O.K.}$$

Anchorage: The *development length*, from Table 6.1, is 53.13 in. for a No. 9 bar, which is available.   The *positive reinforcement* must be taken 6 in. beyond the face of the support.   The *negative reinforcement* must be taken an additional length ℓ_a beyond the point of inflection.   We will take the point of inflection conservatively as $\frac{1}{4}\ell = 7$ ft 6 in. from the face of the support.
The additional length is

$$\ell_a \geq d = 19.33 \text{ in.}$$

$$\geq 12d_b = 13.54 \text{ in. for a No. 9 bar}$$

$$\geq \frac{1}{16}\ell = 1.875 \text{ ft} = 22.5 \text{ in.} = 1 \text{ ft } 10\frac{1}{2} \text{ in.}$$

Therefore, all bars must be taken 7 ft. 6 in. + 1 ft 10½ in. = 9 ft. 4½ in. beyond the face of the support.   The layout of the flexural reinforcement is shown in Fig. 9.10.   The shear reinforcement is designed in Section 11.5.

Example 9.9B

Design a rectangular beam continuous over three equal spans of 10 m, carrying a dead load of 12 kN/m and live load of 7 kN/m, using Grade 25 concrete and Grade 400 steel.

Data given: $D = 12$ kN/m, $L = 7$ kN/m, $\ell = 10$ m (continuous over three span), $f'_c = 25$ MPA, $f_y = 400$ MPa, and $\phi = 0.90$.

Bending movements: Using the theory of continuous beams, the bending moment coefficients, from Table 7.1, are those in Fig. 9.9. The ultimate loads are

$$w_{uD} = 1.4 \times 12 = 16.8 \text{ kN/m}$$

$$w_{uL} = 1.7 \times 7 = 11.9 \text{ kN/m}$$

From Fig. 9.9, the ultimate bending moments are:

	Due to Dead Load	Due to Live Load	Total M_u
Maximum negative	168	139	307 kN·m
Positive, outer span	134	120	254 kN·m
Positive, center span	42	89	131 kN·m

Dimensions of section: We will first determine the effective depth for the maximum bending moment, using Eq. (9.18), which gives $\rho = 0.18 f'_c/f_y = 0.011$, and Table 9.6, which gives $z/d = 0.89$. From Eq. (9.19), the product is

$$bd^2 = \frac{M_u}{\phi f_y \rho} \frac{d}{z}$$

$$= \frac{307 \times 10^{-3}}{0.90 \times 400 \times 0.011 \times 0.89} = 87.1 \times 10^{-3} \text{ m}^3$$

For $b = 0.3$ m, $d = 0.539$ m. Assuming that the largest bar is No. 30, and that No. 10 stirrups are needed, the overall depth is

$$h = 539 + \tfrac{1}{2} \times 30 + 10 + 40 = 604 \text{ mm}$$

Using a 300-mm by 625-mm beam.

Since the actual overall depth is so much larger than the calculated value, we will recalculate the effective depth for steel economy:

$$d = 625 - 40 - 10 - \tfrac{1}{2} \times 30 \text{ (say)} = 560 \text{ mm}$$

The moment arm is

$$z = 0.89 \times 560 = 498 \text{ mm}$$

Reinforcement: The area of reinforcement at each section is

$$A_s = \frac{M_u}{\phi f_y z} = \frac{M_u}{0.90 \times 400 \times 0.498} \text{ m}^2$$

		Reinforcement
Negative	1712 mm²	Use 2 No. 30 + 2 No. 20 = 2000 mm²
Positive, outer span	1416 mm²	Use 2 No. 30 + 2 No. 15 = 1800 mm²
Positive, inner span	731 mm²	Use 4 No. 20 = 1200 mm²

Detailing: The 2 No. 30 bars serve as negative reinforcement and as positive reinforcement in the outer span. The 2 No. 20 bars serve as negative reinforcement and as positive reinforcement in the center span. The center span has 2 No. 20 bars coming from the left and another 2 coming from the right, making 4 bars. This is more than needed, but the bars are available because they are needed over the neighboring supports. The 2 No. 15 bars which provide the additional positive reinforcement in the outer span, are bent up over the outer supports, where a small negative moment should be allowed for, although theoretically there is none. The arrangement is shown in Fig. 9.10 in American units.

The critical combination for *width* are 2 No. 30 + 2 No. 20 bars, which require b = 2 × 40 (cover) + 2 × 10 (ties) + 2 × 30 (bars) + 2 × 20 (bars) + 2 × 30 (spaces between No. 30 and No. 20) + 1 × 20 (space between 2 No. 20) = 280 mm (300 mm). O.K.

Deflection criterion: From Table 8.2, for the outer span, the minimum depth is

$$h = \frac{\ell}{24} = \frac{10\ 000}{24} = 417 \text{ mm (625 mm)} \quad \text{O.K.}$$

EXERCISES

1. What happens when there is a primary tension failure of a reinforced concrete beam or slab? Does the reinforcing steel break in the event of a tension failure?
2. What happens when there is a primary compression failure of a reinforced concrete beam? Why is a primary compression failure less likely to happen in a slab?
3. What is a balanced failure condition?
4. Why do concrete codes not permit designs that would result in a primary compression failure? How do they do that?
5. What is the neutral axis? What determines its depth?
6. What is the effective depth?
7. Is the reinforcement ratio the ratio of the reinforcement to the concrete? If not, what is it?
8. What is the moment arm? How is its length determined?

The answers to the numerical exercises are given in Appendix E.

9. A reinforced concrete slab is continuous over three equal clear spans of 12 ft 6 in , and it carries a dead load of 90 psf and a live load of 60 psf. Determine the maximum bending moment, using the bending moment coefficients of the *ACI Code* (Fig. 7.4).

10. Determine the overall depth required for the slab of Exercise 9 to satisfy the maximum deflection criterion.

11. Determine the amount of main and secondary reinforcement required for the slab of Exercise 10, using Grade 3 concrete and Grade 60 steel, and assuming $z/d = 0.95$.

12. Choose suitable bar sizes and spacing for the reinforcement calculated in Exercise 11.

13. Determine the effective depth and the amount of main reinforcement required for a reinforced concrete slab subjected to a bending moment of 6000 lb · ft/ft, using Grade 4 concrete and Grade 60 steel. Ignore the deflection criterion, and use a reinforcement ratio of 0.006.

14. Determine the overall depth and choose suitable bar sizes and spacing for the slab of Exercise 13.

15. Determine the effective depth and the amount of main reinforcement required for a reinforced concrete slab subject to a bending moment of 80,000 lb · in./ft, using Grade 3.5 concrete and Grade 40 steel. Ignore the deflection criterion, and use a reinforcement ratio of 0.007.

16. Determine the overall depth and choose suitable bar sizes and spacing for the slab of Exercise 15.

17. A reinforced concrete slab, 6½ in. thick, is reinforced with No. 4 bars at 10½-in. centers. The concrete is Grade 3 and the reinforcement is grade 60. What is its maximum ultimate resistance moment?

18. If the slab of Exercise 17 is simply supported over a span of 12 ft 6 in. and the dead load is 90 psf, what is the permissible live load?

19. A reinforced concrete slab, 8 in. thick, is reinforced with No. 4 bars at 7-in. centers. The concrete is Grade 4 and the reinforcement Grade 60. What is the maximum ultimate resistance moment?

20. A reinforced concrete beam, continuous over five equal spans of 32 ft each, carries a uniformly distributed dead load of 900 lb/ft and a live load of 550 lb/ft. Determine the maximum bending moment, using the theory of continuous beams (Table 7.1).

21. Determine a suitable width and effective depth and the amount of reinforcement required for the beam in Exercise 20 if the concrete is Grade 4 and the reinforcement is Grade 60.

22. Choose suitable reinforcing bars for the beam of Exercise 21, and determine the overall depth.

23. A reinforced concrete beam is subject to a maximum bending moment of 300,000 lb · ft. Determine a suitable width and effective depth and the amount of reinforcement required if the concrete is Grade 3 and the reinforcement is Grade 60.

24. A reinforced concrete beam, 15 in. wide and 30 in. deep, is reinforced with 8 No. 7 bars in two layers. Determine the ultimate resistance moment if the concrete is Grade 3.5 and the reinforcement is Grade 60.

25. If the beam of Exercise 24 is simply supported over a span of 32 ft and carries a

uniformly distributed dead load of 1200 lb/ft, determine the permissible uniformly distributed live load.

26. A reinforced concrete beam, continuous over two spans, is 12 in. wide and 30 in. deep, and it is reinforced with 6 No. 7 bars in two layers. Determine its ultimate resistance moment if the concrete is Grade 5 and reinforcement is Grade 60.

Metric Exercises

27. A reinforced concrete slab is continuous over three equal clear spans of 3.75 m, and it carries a dead load of 4.5 kPa and a live load of 2.8 kPa. Determine the maximum bending moment, using the bending moment coefficients of the *ACI Code* (Fig. 7.4).

28. Determine the overall depth required for the slab of Exercise 27 to satisfy the maximum deflection criterion.

29. Determine the amount of main and secondary reinforcement required for the slab of Exercise 28, using Grade 20 concrete and Grade 400 steel, and assuming $z/d = 0.95$.

30. Choose suitable bar sizes and spacing for the reinforcement calculated in Exercise 29.

31. Determine the effective depth and the amount of main reinforcement required for a reinforced concrete slab subjected to a bending moment of 24 kN · m/m, using Grade 30 concrete and Grade 400 steel. Ignore the deflection criterion, and use a reinforcement ratio of 0.006.

32. Determine the overall depth, and choose a suitable bar size and spacing for the slab of Exercise 31.

33. A reinforced concrete slab, 150 mm thick, is reinforced with No. 15 bars at 200-mm centers. The concrete is Grade 25 and the reinforcement is Grade 400. What is the maximum ultimate resistance moment?

34. If a slab is simply supported over a span of 4.5 m and the dead load is 4 kPa, what is the permissible live load?

35. A reinforced concrete slab, 400 mm thick, is reinforced with No. 20 bars at 200-mm centers. The concrete is Grade 30 and the reinforcement Grade 400. What is the maximum ultimate resistance moment?

36. A reinforced concrete beam continuous over five equal spans of 10.5 m each carries a uniformly distributed dead load of 10 kN/m and a live load of 7.5 kN/m. Determine the maximum bending moment using the theory of continuous beams (Table 7.1).

37. Determine a suitable width and effective depth and the amount of reinforcement required for the beam in Exercise 36 if the concrete is Grade 25 and the reinforcement Grade 400.

38. Choose suitable reinforcing bars for the beam of Exercise 37 and determine the overall depth if the concrete is Grade 25 and the reinforcement is Grade 400.

39. A reinforced concrete beam is subject to a maximum bending moment of 400 kN · m. Determine a suitable width and effective depth and the amount of

reinforcement required if the concrete is Grade 20 and the reinforcement is Grade 400.

40. A reinforced concrete beam, 375 mm wide and 750 mm deep, is reinforced with 8 No. 20 bars in two layers. Determine the ultimate resistance moment if the concrete is Grade 20 and the reinforcement is Grade 400.

41. If the beam of Exercise 40 is simply supported over a span of 9 m and carries a uniformly distributed dead load of 16 kN/m, determine the permissible uniformly distributed live load.

42. A reinforced concrete beam, continuous over two spans, is 300 mm wide and 750 mm deep and is reinforced with 6 No. 20 bars in two layers. Determine its ultimate resistance moment if the concrete is Grade 30 and the reinforcement is Grade 400.

Chapter 10

The Monolithic Construction of Beams and Slabs

This chapter deals with beams that are cast monolithically with the slabs, that is, the beams supporting one-way slabs (Chapter 9) and two-way slabs (Chapter 12). Some of these beams are T-beams at midspan, and some require compression reinforcement near the supports.

10.1 DEFINITION OF T-BEAMS AND L-BEAMS

A reinforced concrete floor system of slabs supported on beams is almost invariably cast monolithically. Moreover, the slab reinforcement runs over the beam, and the stirrups in the beam project into the slab. The beam and the adjoining portion of the slab thus deform like one unit, and the same stresses exist in the beam and in the adjoining portion of the slab. It is consequently admissible to treat the slab as a part of the beam which may be regarded as T-shaped, or as L-shaped in the case of a beam at the end of a floor or roof structure.

The slab thus performs a dual function: it spans as a slab between the supporting webs, and it forms part of the beam at right angles (Fig. 10.1). However, since the stresses due to the slab action are at right angles to stresses due to the beam action, neither influences the other. (The theory of the composition of stresses is explained in most elementary textbooks on the theory of structures, e.g., Ref. 7.3)

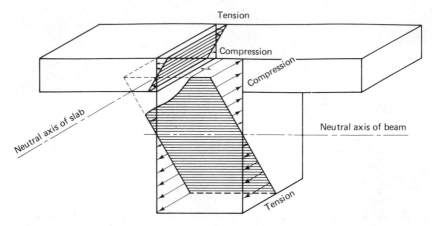

Figure 10.1   Strain distribution in a T-beam and in the slab spanning at right angles to the beam.

This composite action of the beam and the slabs ("T-beam action") was already appreciated in the nineteenth century, when it was noticed that such floors were much stiffer than the beams alone, without the floor slabs.

The stress in the portion of the flange over the web is higher than in the outstanding portions of the flange (Fig. 10.2), but it is customary to assume uniform stress distribution across the effective width of the flange.

The effective width of the flange must be restricted when the webs are relatively widely spaced, since the stress in the flange is reduced with distance from the web. Theoretical solutions for the effective width of T-beams based on elastic theory have been obtained, but it is usually determined by simple empirical rules (Fig. 10.3).

When the webs are closely spaced and the neutral axis is near the base of the flange, the entire floor has virtually the same strength as a solid slab with the same effective depth and the same reinforcement (Fig. 10.4). Since the concrete below the neutral axis is in tension and does not contribute to

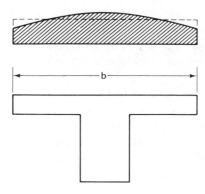

Figure 10.2   Approximate distribution of compressive stress at the top of the flange of a T-beam. The flange is the portion of the slab included in the beam, and it stretches right across the beam for the full width b. The web is the portion of the beam *below* the flange only.

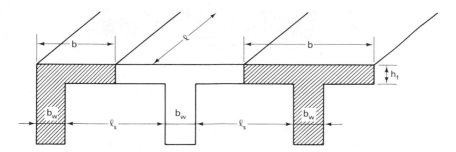

Figure 10.3   Effective width of the flanges of T-beams and L-beams in accordance with Section 8.10 of the *ACI Code*.   L-beam: $b \leq b_w + \frac{1}{12}\ell$; $\leq b_w + 6h_f$; $\leq b_w + \frac{1}{2}\ell_s$.   T-beam: $b \leq \frac{1}{4}\ell \leq b_w + 16h_f$; $\leq b_w + \ell_s$.

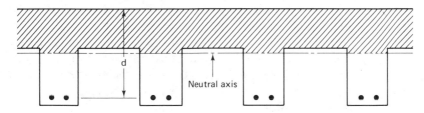

Figure 10.4   T-beam with closely spaced webs, or ribs.

the flexural resistance, it can be cut away in part, and the reinforcement concentrated in the webs.   The T-beam system may then be regarded as a solid slab from which part of the concrete has been removed to reduce the weight.   It should be noted, however, that while the T-beam system has almost the same strength as the solid slab, it has a larger deflection.

A T-beam, as defined by the *ACI Code*, does not necessarily look like a T-beam (Fig. 10.5).   On the other hand, a beam that looks like a T-beam is not necessarily a T-beam as defined above (Fig. 10.6).

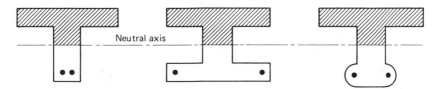

Figure 10.5   A T-beam does not necessarily have to look like a T-beam.   These precast T-beams have identical strength, but look very different.   The resistance moment of a T-beam depends only on the shape of the compression flange and web, and on the area and location of the reinforcement.   The shape of the concrete on the tension side of the neutral axis does not affect it.

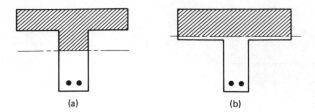

Figure 10.6   A beam that looks like a T-beam is not necessarily a T-beam.   (a) The neutral axis falls within the web, and the beam is thus a T-beam.   (b) The neutral axis falls within the flange, and the shape of the concrete section above the neutral axis is rectangular.   This beam is a rectangular beam.

10.2 USE OF COMPRESSION REINFORCEMENT

The two main uses of compression reinforcement are to increase the stiffness of the section and to increase the strength.

Compression reinforcement increases stiffness more than strength, and it thus enables us to reduce the overall depth if the deflection is calculated from the moment of inertia (Table 8.1).   Moreover, the presence of compression steel reduces creep, and this is allowed for in the assessment of long-term deflection (Section 8.2).

In this chapter, however, we are concerned with the effect of the compression reinforcement on strength.   Continuous concrete beams are usually cast monolithically with the concrete slab.   At midspan the bending moments are positive, and the compression face of the beam merges with the slab, which thus becomes a part of the beam (see Section 10.1).   At the support section the bending moments are negative, and the slab is thus on the tension face of the beam, where it cannot contribute to its strength (Fig. 10.7).   We thus have a change from a width of beam b at midspan [Fig. 10.7(a)] to a much smaller width of beam b_w [Fig. 10.7(b)] at the support.

This does not necessarily reduce the strength of the beam, if the reinforcement ratio ρ is low, so that the concrete is understressed.   However, as

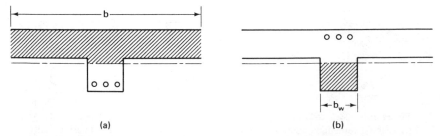

Figure 10.7   The same beam behaves very differently at midspan (a) and at the support (b).   At the midspan section the bending moment is positive, the compression face is on top, and its width is b.   At the support section the bending moment is negative, the compression face is at the bottom, and its width is b_w.

the reinforcement ratio ρ increases, the change in the concrete width from b to b_w (i.e., the loss of the concrete in the overhangs of the T-beam flange) may change a primary tension failure into a primary compression failure, and this is not admissible (see Section 9.3).

We must then add an "invisible" flange to the web of the beam to restore a primary tension failure. Since steel is 10 to 20 times as strong as concrete, we greatly increase the compressive force in the web by replacing some of the concrete in compression with the much stronger compression steel. The external appearance of the beam is not affected.

There are two other reasons for using compression steel, although neither is formally admitted by the *ACI Code*. Our understanding of structural mechanics is not as perfect as we would like it to be, and for very complex structures the sign of the bending moment cannot always be predicted with complete certainty. If the "tension" reinforcement is in fact on the compression face, and there is no reinforcement on the actual tension face, the structure will be seriously damaged and may even collapse. By having reinforcement on both sides we do not have to be quite so certain of the sign of the bending moment provided we can make a reasonable prediction of the magnitude of the moment.

Another argument in favor of compression steel is its effect on the ultimate strength of the frames. Although steel codes permit the use of the plastic theory for the design of rigid *frames* under certain conditions, concrete codes do not.[1] The forces and moments in rigid concrete frames are calculated by the elastic theory, and only the dimensions of the cross section are determined by ultimate strength design (see Section 7.2). Nevertheless, some plastic redistribution occurs in all reinforced concrete frames, and this is greatly improved by the insertion of compression reinforcement. The formation of a plastic hinge in reinforced concrete is a function of the steel: the steel is the plastic material; the concrete is brittle. If there is steel on both faces of the concrete, the plasticity of the hinge is increased, and so is the plastic moment redistribution at the ultimate load. The frame thus has a greater capacity to deform before collapse, and its ultimate load may also be higher. This is, however, merely an additional safeguard which cannot at present be allowed in design according to the *ACI Code*.

Compression reinforcement requires ties to prevent buckling of the bars (see Section 6.6).

10.3 THEORY OF T-BEAMS AND L-BEAMS

The theory of an L-beam (i.e., a flanged beam at the end of the slab) is precisely the same as that of a T-beam, except that the width b is less (Fig. 10.3).

---

[1]The *ACI code*, Section 8.4, does, however, permit moment redistribution in continuous *beams*. This is the reason for the moment adjustment shown in Fig. 7.2.

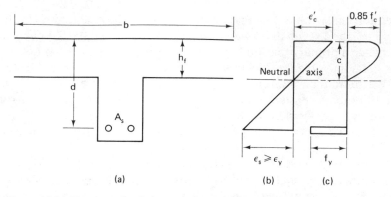

Figure 10.8 The neutral axis is precisely at the base of the flange of a T-beam, the borderline case between a T-beam and a rectangular beam. (a) Dimensions of the section. (b) Distribution of strain. (c) Distribution of stress.

Let us consider a T-beam with a flange of width b and thickness h_f and reinforcement A_s at an effective depth d [Fig. 10.8(a)]. The neutral axis could be either within the flange, or below the flange, or for the limiting case at the base of the flange. Let us consider that limiting case, when the strain ranges from the ultimate compressive ϵ_c' on top, to a steel strain ϵ_s at the bottom which is either equal or greater than the yield or proof strain ϵ_y, so that the steel has yielded and the stress in it is f_y. The strain changes from tension to compression at the neutral axis, and its depth

$$c = h_f \tag{10.1}$$

If $c < h_f$, the neutral axis lies within the flange, and we have a rectangular beam of width b, reinforced with an area of tension steel A_s at an effective depth d (Fig. 10.9). Its ultimate resistance moment, from Eq. (9.8), is

$$M' = \phi f_y A_s z \tag{10.2}$$

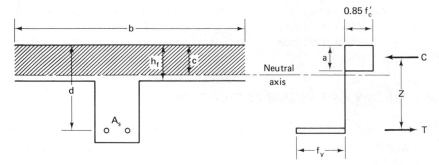

Figure 10.9 Resistance moment of a T-beam (actually a rectangular beam) when the neutral axis falls within the flange.

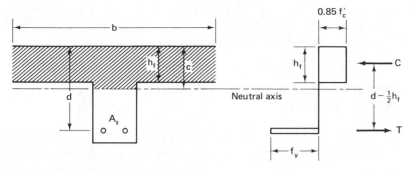

Figure 10.10   Resistance moment of a T-beam when the neutral axis is below the flange.

where z/d has the value listed in Table 9.1, or it may be taken approximately as $z/d = 0.95$.

If $c > h_f$, the neutral axis lies within the web, and we have a flange of width b and depth h_f, stressed at failure to $0.85f_c'$, and an area of reinforcement A_s at an effective depth d. These form a resistance moment (Fig. 10.10), with a moment arm $d - \frac{1}{2}h_f$. Consequently,

$$M' = \phi f_y A_s (d - \tfrac{1}{2}h_f) \tag{10.3}$$

The limiting condition occurs when the neutral axis is just at the base of the flange, as in Fig. 10.8. Let us express the resistance moment in terms of the compressive force:

$$M' = \phi C z$$

Substituting from Eq. (9.2),

$$M' = \phi 0.85 f_c' b \beta_1 c z$$

Putting $\phi = 0.90$, $\beta_1 = 0.85$ (approximately), and $z = 0.90d$ (approximately), the beam is rectangular if

$$\frac{h_f}{d} \geqslant \frac{c}{d} = \frac{1.71M'}{f_c' b d^2} \tag{10.4}$$

Equations (10.2) and (10.3) hold true until we reach the limiting reinforcement ratio, ρ_{max}, listed in Table 9.4. However, this is extremely unlikely, since $A_s = \rho_{max} bd$ represents so large an amount of reinforcement that it could not easily be accommodated within the web. It is also improbable that such a beam could pass the deflection criterion (Section 8.7).

Since the overall depth of T-beams and the thickness of their flanges are usually known before the amount of reinforcement is calculated, we can determine from Eq. (10.4) whether the neutral axis falls inside or outside the flange, and consequently whether Eq. (10.2) or Eq. (10.3) should be used.

Equation (10.4) contains some approximations, so that it is not entirely accurate when the neutral axis is near the bottom of the flange, but when this is so, Eqs. (10.2) and (10.3) give the same numerical answer.

10.4 THEORY OF BEAMS WITH COMPRESSION REINFORCEMENT

Let us consider a rectangular beam of width b, which has a cross-sectional area A_s of tension reinforcement at an effective depth d, and a cross-sectional area of compression reinforcement A'_s at a depth d' from the compression face (Fig. (10.11)).

We will divide this section into two parts, one containing all the concrete and the greatest permissible area of tension reinforcement, and the second containing no concrete, but all the compression steel and the remainder of the tension reinforcement.

The resistance moment of the first part is, from Section 9.4,

$$M_1 = M'_{max} = k_m b d^2 \qquad (10.5)$$

and the corresponding area of tension reinforcement

$$A_{s1} = \rho_{max} b d \qquad (10.6)$$

where k_m and ρ_{max} have the values listed in Table 9.5.

The remainder of the ultimate resistance moment

$$M_2 = M_u - M_1 \qquad (10.7)$$

is resisted by the compression reinforcement A'_s and the remainder of the

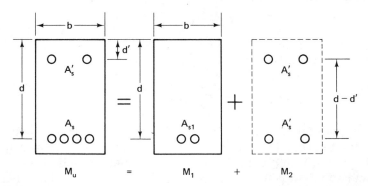

Figure 10.11 We analyze a beam with compression reinforcement by dividing its ultimate resistance moment M_u into two parts M_1 and M_2. The resistance moment M_1 is provided by the entire concrete section and the greatest possible area of tension reinforcement A_{s1} permissible for a rectangular section. The resistance moment M_2 is provided by the remainder of the tension reinforcement A_{s2} and the entire compression reinforcement A'_s.

tension reinforcement A_{s2}. The length of the moment arm is $d - d'$. Consequently,

$$A_{s2} = \frac{M_2}{f_y(d - d')} \tag{10.8}$$

and the area of tension reinforcement is

$$A_s = A_{s1} + A_{s2} \tag{10.9}$$

The area of compression reinforcement is

$$A'_s = A_{s2} \tag{10.10}$$

provided that the compression steel reaches the yield point before the bending moment reaches the ultimate resistance moment M_u. If it does not, then

$$A'_s = \frac{A_{s2}f_y}{f'_s} \tag{10.11}$$

where f'_s is the actual stress in the compression reinforcement at failure.

10.5 LIMITATION ON THE STRESS IN THE COMPRESSION REINFORCEMENT

Since we assume that plane sections remain plane, the strain in the compression reinforcement (Fig. 10.12)

$$\epsilon'_s = \epsilon'_c \frac{c - d'}{c} \tag{10.12}$$

where ϵ'_c is the strain in the concrete at failure, which the *ACI Code* takes as 0.003 for all values of the concrete strength (see Fig. 8.4), and c is the depth of the neutral axis at failure. Since we are taking the netural axis as

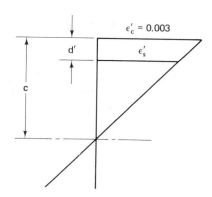

Figure 10.12 We have assumed in Section 8.5 that plane sections remain plane, so that strains are proportional to the distance from the neutral axis.

low as we are permitted, to get the greatest possible contribution from the concrete, c has the values c_{max} calculated in Table 9.4.

The stress in the compression reinforcement at failure is

$$f'_s = E_s\epsilon'_s$$

where $E_s = 29,000,000$ psi, is the modulus of elasticity of steel (see Section 6.3). However, this applies only when $f'_s \leqslant f_y$. After the stress in the compression steel has reached f_y the strain in the compression steel increases further without increase in the stress (see Fig. 8.5). Consequently,

$$f'_s = E_s\epsilon'_s = E_s\epsilon'_c \left[1 - \frac{d'}{c_{max}}\right] \leqslant f_y \qquad (10.13)$$

Substituting from Figs. 8.4 and 8.6 and from Table 9.4,

$$E_s = 29,000,000 \text{ psi}; \quad \epsilon'_c = 0.003; \quad \frac{c_{max}}{d} = 0.514 \text{ for}$$

$$f_y = 40,000 \text{ psi}, \quad \text{and} \quad \frac{c_{max}}{d} = 0.444 \text{ for } f_y = 60,000 \text{ psi}$$

we obtain that $f'_s = f_y$ if

$$d' \leqslant 0.28d \quad \text{for} \quad f_y = 40,000 \text{ psi} \qquad (10.14a)$$

and

$$d' \leqslant 0.14d \quad \text{for} \quad f_y = 60,000 \text{ psi} \qquad (10.15a)$$

In metric units, $f'_s = f_y$ if

$$d' \leqslant 0.25d \quad \text{for} \quad f_y = 300 \text{ MPa} \qquad (10.14b)$$

and

$$d' \leqslant 0.15d \quad \text{for} \quad f_y = 400 \text{ MPa} \qquad (10.15b)$$

10.6 USE OF COMPRESSION REINFORCEMENT FOR SECTIONS OF LIMITED DEPTH

For architectural reasons beams must sometimes be limited in depth. When the reinforcement ratio ρ_{max} (see Table 9.4) is reached, a further reduction in depth can be achieved only by increasing the width, but since the strength is proportional to bd^2, this leads to an increase in the amount of concrete. With compression reinforcement depth is further reduced by up to 30%.

Compression reinforcement is also useful for limiting deflection, a problem associated with shallow concrete sections. The long-term deflection is due primarily to creep caused by the squeezing of water from the pores of concrete by sustained loads. Steel is not similarly affected, and therefore restrains creep.   Hence compression reinforcement reduces the long-term deflection (see Section 8.2).

10.7 USE OF COMPRESSION REINFORCEMENT IN SEISMIC DESIGN

Concrete is a brittle material and is liable to crack when subjected to a sudden earthquake force.   The ductility of reinforced concrete is due entirely to the reinforcing steel.   In seismic design (see also Sections 9.3 and 13.10) compression reinforcement is helpful for two reasons:

1. The earthquake forces may cause a reversal of bending moment; if the unreinforced compression face becomes the tension face, even momentarily, the structure may be seriously damaged.
2. Even if no moment reversal occurs, the addition of a ductile material on the compression face improves the ductility of the structure.   A section with approximately the same amount of compression and tension reinforced and adequate shear reinforcement behaves like a section with two steel flanges and a steel web encased in concrete.

10.8 DESIGN EXAMPLES

T-beams and compression reinforcement formed a part of normal construction prior to the introduction of ultimate strength design, tentatively in the *ACI Code* of 1956, and definitely in the *ACI Code* of 1971.   This method, described in Chapter 9, allows much higher percentages of reinforcement for balanced design than the older service-load method mentioned in Section 8.1, and thus reduces the need for compression reinforcement.   It also results in a much smaller neutral-axis depth ratio at failure than is obtained under service loads. Therefore, most of the "T-beams" in small building frames are in fact rectangular beams, and compression reinforcement is rarely required in small reinforced concrete buildings.

The loads and spans in Chapter 9 are realistic, but higher loads are needed to obtain T-beams and beams with compression steel.   The following examples have therefore been chosen with bending moments that illustrate the transition from rectangular beam to T-beam, and the transition from tension reinforcement only to tension and compression reinforcement.

Example 10.1A

A reinforced concrete beam has a flange 3 ft wide and 4 in. thick, a web 12 in. wide, and an effective depth of 21 in.   Design the reinforcement required for an ultimate positive bending moment of 300,000 lb·ft. using Grade 4 concrete and Grade 60 steel.

Data given: $b = 36$ in., $b_w = 12$ in., $h_f = 4$ in., $d = 21$ in., $M_u = 3,600,000$ lb·in., $f_c' = 4000$ psi, $f_y = 60,000$ psi, and $\phi = 0.90$.
The ratio

$$\frac{h_f}{d} = \frac{4}{21} = 0.190$$

From Eq. (10.4),

$$\frac{1.71 \times 3,600,000}{4000 \times 36 \times 21^2} = 0.097 < \frac{h_f}{d}$$

The beam is therefore a rectangular beam, and from Eq. (10.2), assuming that $z = 0.95d$,

$$A_s = \frac{3,600,000}{0.90 \times 60,000 \times 0.95 \times 21} = 3.342 \text{ in.}^2$$

In a web 12 in. wide we can accommodate 4 No. 8 bars, but only 3 bars of a larger size. Assuming that there is an adequate length available for anchoring the bars. *Use 3 No. 10 bars* ($= 3.81$ in.^{2}).

Only nominal stirrups are needed in the region of the positive reinforcement (see Section 11.5). These stirrups are normally bent around two small (i.e., No. 3) longitudinal bars. Overall depth is

$$h = 21 + \tfrac{1}{2} \times 1.27 + 0.38 + 1.5 = 23.5 \text{ in.}$$

Use a beam 12 in. wide by 24 in. deep.
The reinforcement ratio is

$$\rho = \frac{3.81}{36 \times 24} = 0.004\ 4$$

so that, from Table 9.4, the maximum concrete stress need not be checked. The layout of the reinforcement is shown in Fig. 10.13.

Example 10.1B

A reinforced concrete beam has a flange 1 000 mm wide and 100 mm thick, a web 300 mm wide, and an effective depth of 525 mm. Design the reinforcement required for an ultimate positive bending moment of 500 kN·m, using Grade 25 concrete and Grade 400 steel.

Data given: $b = 1$ m, $b_w = 0.30$ m, $h_f = 0.1$ m, $d = 0.525$ m, $M_u = 500 \times 10^{-3}$ MN·m, $f_c' = 25$ MPa, $f_y = 400$ MPa, and $\phi = 0.90$.
The ratio

$$\frac{h_f}{d} = \frac{0.1}{0.525} = 0.190$$

From Eq. (10.4),

$$\frac{1.71 \times 500 \times 10^{-3}}{25 \times 1 \times 0.525^2} = 0.124 < \frac{h_f}{d}$$

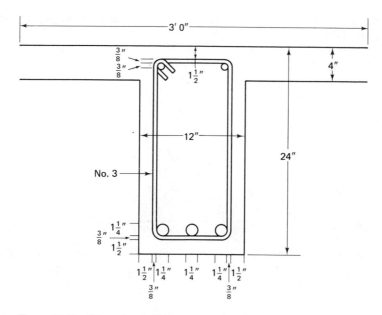

Figure 10.13 Dimensions of the cross section designed in Example 10.1.

The beam is therefore a rectangular beam, and from Eq. (10.2), assuming that $z = 0.95d$,

$$A_s = \frac{500 \times 10^{-3}}{0.90 \times 400 \times 0.95 \times 0.525} = 2.785 \times 10^{-3} \text{ m}^2 = 2\,785 \text{ mm}^2$$

In a web 300 mm wide we can accommodate 4 No. 25 bars, but only 3 bars of a larger size. Assuming that there is an adequate length available for anchoring the bars. *Use 3 No. 35 bars* ($= 3\,000$ mm²).

Only nominal stirrups are needed in the region of the positive reinforcement (see Section 11.5). These stirrups are normally bent around two small (i.e., No. 10) longitudinal bars. Overall depth is

$$h = 525 + \frac{1}{2} \times 35 + 10 + 40 = 593 \text{ mm}$$

Use a beam 300 mm wide by 600 mm deep.
The reinforcement ratio is

$$\rho = \frac{3\,000}{1\,000 \times 525} = 0.005\,7$$

so that, from Table 9.4, the maximum concrete stress need not be checked. The layout of the reinforcement is shown in Fig. 10.13 in U.S. units.

Example 10.2A

Design the reinforcement for a concrete beam with a flange 16 in. wide and 3 in. thick. The remaining particulars are the same as for Example 10.1.A.

Data given: b = 16 in., b_w = 12 in., h_f = 3 in., d = 21 in., M_u = 3,600,000 lb·in., f'_c = 4,000 psi, f_y = 60,000 psi, and ϕ = 0.90.
   The ratio

$$\frac{h_f}{d} = 0.143$$

From Eq. (10.4),

$$\frac{1.71 \times 3,6000,000}{4000 \times 16 \times 21^2} = 0.218 > \frac{h_f}{d}$$

The beam is therefore a T-beam, and, from Eq. (10.3),

$$A_s = \frac{3,600,000}{0.90 \times 60,000(21 - \frac{1}{2} \times 3)} = 3.419 \text{ in.}^2$$

As one would expect this value is slightly higher than that calculated in Example 10.1.A. *Use 3 No. 10 bars* (= 3.81 in.^{2}).
   The reinforcement ratio is

$$\rho = \frac{3.81}{16 \times 21} = 0.011$$

so that the maximum concrete stress (Table 9.5), although higher than in Example 10.1.A, need not be checked. The reinforcement is the same as for Example 10.1.A, sketched in Fig. 10.13.

Example 10.2B

Design the reinforcement for a concrete beam with a flange 500 mm wide and 75 mm thick. The remaining particulars are the same as for Example 10.1.B.

Data given: b = 0.5 m, b_w = 0.30 m, h_f = 0.075 m, d = 0.525 m, M_u = 500 × 10^{-3} MN·m, f'_c = 25 MPa, f_y = 400 MPa, and ϕ = 0.90.
   The ratio

$$\frac{h_f}{d} = 0.143$$

From Eq. (10.4),

$$\frac{1.71 \times 500 \times 10^{-3}}{25 \times 0.5 \times 0.525^2} = 0.248 > \frac{h_f}{d}$$

The beam is therefore a T-beam and, from Eq. (10.3),

$$A_s = \frac{500 \times 10^{-3}}{0.90 \times 400 \times (0.525 - \frac{1}{2} \times 0.075)} = 2.849 \times 10^{-3} \text{ m}^2 = 2\,849 \text{ mm}^2$$

As one would expect, this value is slightly higher than that calculated in Example 10.1.B. *Use 3 No. 35 bars* (= 3 000 mm^2).

The reinforcement ratio is

$$\rho = \frac{3\,000}{500 \times 525} = 0.011$$

so that the maximum concrete stress, although higher than in Example 10.1.B, need not be checked.

The reinforcement is the same as for Example 10.1.B.

Example 10.3A

Design the reinforcement for a concrete beam with the same dimensions as that of Example 10.1.A, but subject to a negative bending moment of 300,000 lb·ft, using Grade 4 concrete and Grade 60 steel.

Since the beam is subject to a negative moment, the compression face is at the base of the web, and the flange is in tension, and thus unable to make the special contribution on which the resistance moments on the beam in Example 10.1.A and 10.2.A depend. More reinforcement is needed, and we will therefore make the effective depth 20 in. to produce an overall depth of 24 in.

Data given: b = 12 in., d = 20 in., M_u = 3,600,000 lb·in., f_c' = 4000 psi, f_y = 60,000 psi, ϕ = 0.90, k_m = 881 psi, and z = 0.807d (from Table 9.5).
From Eq. (10.5),

$$M_1 = k_m bd^2 = 881 \times 12 \times 20^2 = 4,229,000 \text{ lb·in.}$$

This is more than the ultimate bending moment M_u, and therefore compression reinforcement is not required. The tension reinforcement, from Eq. (9.8), is

$$A_s = \frac{M_u}{\phi f_y z} = \frac{3,600,000}{0.90 \times 60,000 \times 0.807 \times 20} = 4.131 \text{ in.}^2$$

If arranged in one layer, this would require 3 No. 11 bars ($= 4.68$ in.²), which is wasteful and is also likely to present anchorage problems.

The beam needs No. 3 stirrups for shear reinforcement. Their spacing is considered in Section 11.5. A 12-in. web can therefore accommodate 4 bars in one layer if their size does not exceed No. 8:

$$b = 2 \times 1.5 \text{ (cover)} + 2 \times 0.38 \text{ (stirrup)} + 4 \times 1.0 \text{ (bars)} + 3$$

$$\times 1.0 \text{ (spaces between bars)} = 10.75 \text{ in.}$$

Use 4 No. 7 + 4 No. 6 ($= 4.16$ in.²).

The bars are separated by a few No. 8 spacer bars (see Section 6.4), and the effective depth is measured to the center of these bars. Hence the overall depth is

$$h = 20 + \tfrac{1}{2} \times 1.0 + 0.88 + 0.38 + 1.5 = 23.26 \text{ in.}$$

Use a beam 12 in. wide by 24 in. deep.

The cross section of the beam is sketched in Fig. 10.14.

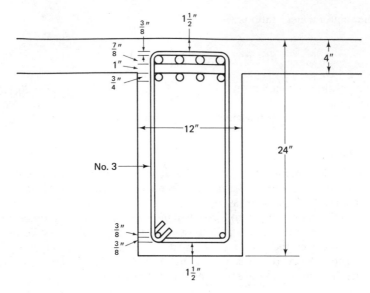

Figure 10.14 Dimensions of the cross section designed in Example 10.3.A.

Example 10.3B

Design the reinforcement for a concrete beam with the same dimensions as that of Example 10.1, but subject to a negative bending moment of 480 kN·m, using Grade 25 concrete and Grade 400 steel.

Since the beam is subject to a negative moment, the compression face is at the base of the web, and the flange is in tension, and thus unable to make the special contribution on which the resistance moments on the beams in Examples 10.1 and 10.2 depend.

Data given: $b = 0.30$ m, $d = 0.525$ m, $M_u = 480 \times 10^{-3}$ MN·m, $f'_c = 25$ MPa, $f_y = 400$ MPa, $\phi = 0.90$, $k_m = 5.92$ MPa, and $z = 0.809d$ (from Table 9.5). From Eq. (10.5),

$$M_1 = k_m bd^2 = 5.92 \times 0.30 \times 0.525^2 = 490 \times 10^{-3} \text{ MN·m}$$

This is more than the ultimate bending moment, M_u, and therefore compression reinforcement is not required. The tension reinforcement, from Eq. (9.8), is

$$A_s = \frac{M_u}{\phi f_y z} = \frac{480 \times 10^{-3}}{0.90 \times 400 \times 0.809 \times 0.525} = 3.139 \times 10^{-3} \text{ m}^2 = 3\,139 \text{ mm}^2$$

If arranged in one layer, this would require 3 No. 45 bars ($= 4\,500$ mm²), which is wasteful and is also likely to present anchorage problems.

The beam needs No. 10 stirrups for shear reinforcement. Their spacing is considered in Section 11.5. A 300-mm web can therefore accommodate 4 bars in one layer if their size does not exceed No. 25:

$$b = 2 \times 40 \text{ (cover)} + 2 \times 10 \text{ (stirrup)} + 4 \times 25 \text{ (bars)} + 3$$

$$\times \ 25 \text{ (spaces between bars)} = 275 \text{ mm}$$

Use 4 No. 25 + 4 No. 20 (= 3 200 mm².)
    The bars are separated by a few No. 25 mm spacer bars (see Section 6.4), and the effective depth is measured to the center of these bars. Hence the overall depth is

$$h = 525 + \tfrac{1}{2} \times 25 + 25 + 10 + 40 = 613 \text{ mm}$$

Use a beam 300 mm wide by 625 mm deep.
    The cross section of the beam is sketched in Fig. 10.14 in American units.

Example 10.4A

Design the reinforcement for the same beam as in Example 10.3.A, but with a negative bending moment of 450,000 lb·ft.
    Data given: $b = 12$ in., $d = 20$ in., $M_u = 5,400,000$ lb·in., $f'_c = 4000$ psi, $f_y = 60,000$ psi, $\phi = 0.90$, $k_m = 881$ psi, and $\rho_{max} = 0.0214$ (from Table 9.5).
    From Eq. (10.5),

$$M_1 = k_m b d^2 = 4,229,000 \text{ lb·in.}$$

This is less than the ultimate bending moment M_u, and therefore compression reinforcement is required for the balance of the ultimate bending moment:

$$M_2 = M_u - M_1 = 5,400,000 - 4,229,000 = 1,171,000 \text{ lb·in.}$$

Assuming one row of No. 7 bars for the compression reinforcement, and No. 3 stirrups, its depth is

$$d' = 1.5 \text{ (cover)} + 0.38 \text{ (stirrup)} + \tfrac{1}{2} \times 0.88 \text{ (bar)} = 2.32 \text{ in.}$$

and the length of the moment arm is

$$d - d' = 20 - 2.32 = 17.68 \text{ in.}$$

The tension reinforcement required for the bending moment M_1 [Eq. (10.6)] is

$$A_{s1} = 0.0214 \times 12 \times 20 = 5.136 \text{ in.}^2$$

and the tension reinforcement required for the moment M_2 [from Eq. (10.8)] is

$$A_{s2} = \frac{1,171,000}{60,000 \times 17.68} = 1.104 \text{ in.}^2$$

Thus the total area of tension reinforcement [Eq. (10.9)] is

$$A_s = 5.136 + 1.104 = 6.240 \text{ in.}^2$$

and the area of compression reinforcement [from Eq. (10.10)] is

$$A'_s = A_{s2} = 1.104 \text{ in.}^2$$

Use 8 No. 8 in two layers for tension reinforcement (= 6.32 in.²) *and 2 No. 7 for compression reinforcement* (= 1.20 in.²).

No. 3 bars are needed as ties for the compression reinforcement, and also as shear reinforcement (see Section 11.5). The shear requirement determines the spacing. The overall depth is the same as for Example 10.3, and the cross section of the beam is sketched in Fig. 10.15.

Example 10.4B

Design the reinforcement for the same beam as in Example 10.3.B, but with a negative bending moment of 600 kN·m.

Data given: b = 0.30 mm, d = 0.525 m, M_u = 600 × 10⁻³ MN·m, f'_c = 25 MPa, f_y = 400 MPa, ϕ = 0.90, k_m = 5.92 MPa, and ρ_{max} = 0.0203 (from Table 9.5). From Eq. (10.5),

$$M_1 = k_m bd^2 = 490 \times 10^{-3} \text{ MN·m}$$

This is less than the ultimate bending moment M_u, and therefore compression reinforcement is required for the balance of the ultimate bending moment:

$$M_2 = M_u - M_1 = (600 - 490) \times 10^{-3} = 110 \times 10^{-3} \text{ MN·m}$$

Assuming one row of No. 20 bars for the compression reinforcement, and 10 mm stirrups, its depth is

$$d' = 40 \text{ (cover)} + 10 \text{ (stirrup)} + \tfrac{1}{2} \times 20 \text{ (bar)} = 60 \text{ mm}$$

and the length of the moment arm is

$$d - d' = 525 - 60 = 465 \text{ mm}$$

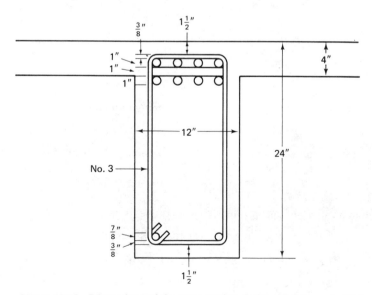

Figure 10.15 Dimensions of the cross section designed in Example 10.4.

The tension reinforcement required for the bending moment M_1 [Eq. (10.6)] is

$$A_{s1} = 0.0203 \times 300 \times 525 = 3\ 197 \text{ mm}^2$$

and the tension reinforcement required for the moment M_2 [from Eq. (10.8)] is

$$A_{s2} = \frac{110 \times 10^{-3}}{400 \times 0.465} = 0.591 \times 10^{-3} \text{ m}^2 = 591 \text{ mm}^2$$

Thus the total area of tension reinforcement [Eq. (10.9)] is

$$A_s = 3\ 197 + 591 = 3\ 788 \text{ mm}^2$$

and the area of compression reinforcement [from Eq. (10.10)] is

$$A'_s = A_{s2} = 591 \text{ mm}^2$$

Use 8 No. 25 in two layers for tension reinforcement (= 4 000 mm²) *and 2 No. 20 for compression reinforcement* (= 600 mm²).

No. 10 bars are needed as ties for the compression reinforcement, and also as shear reinforcement (see Section 11.5). The shear requirement determines the spacing. The overall depth is the same as for Example 10.3.B, and the cross section of the beam is sketched in Figure 10.15 in American units.

EXERCISES

1. Why does a T-beam, as the term is used in reinforced concrete design, not necessarily look like a T-beam? On the other hand, why is a beam that looks like a T not necessarily a T-beam?
2. What is the difference between a T-beam and an L-beam?
3. How wide is the flange of a T-beam?
4. How is the flange of a T-beam, spanning between adjacent webs, designed as a reinforced concrete slab?
5. Reinforcing bars are long, very thin bars that would easily buckle if used in compression. Why do they not buckle in a properly designed reinforced concrete structure?
6. Why does compression reinforcement reduce long-term deflection?
7. Why is the use of compression reinforcement particularly helpful in regions that are subject to earthquakes?
8. The most convenient method for the design of beams with compression reinforcement is to divide them into two parts: one part contains all the concrete, and enough tension reinforcement to produce a balanced design; the other contains the remainder of the reinforcement, but no concrete. Derive expressions for the two component moments and their respective reinforcement.
9. Why does the use of compression reinforcement make it possible to reduce the overall depth of the reinforced concrete structure (as compared to a reinforced concrete structure serving the same purpose that does not employ compression reinforcement)?
10. A reinforced concrete beam has a flange 18 in. wide and 3 in. thick, a web 10 in.

wide, and an effective depth of 20 in.   Determine the reinforcement required for an ultimate positive bending moment of 3,000,000 lb·ft, using Grade 3.5 concrete and Grade 60 steel.

11. Determine the amount of tension and compression reinforcement (if any) required for a beam with a flange 18 in. wide and 3 in. thick, a web 10 in. wide, and an effective depth of 18 in. to resist an ultimate negative bending moment of 3,000,000 lb·ft, using Grade 3.5 concrete and Grade 60 steel.

Metric Exercises

12. A reinforced concrete beam has a flange 450 mm wide and 75 mm thick, a web 250 mm wide, and an effective depth 500 mm.   Determine the reinforcement required for an ultimate positive bending moment of 400 kN·m, using Grade 20 concrete and Grade 400 steel.

13. Determine the amount of tension and compression reinforcement (if any) required for a beam with a flange 400 mm wide and 75 mm thick, a web 250 mm wide, and an effective depth of 450 mm to resist an ultimate negative bending moment of 300 kN·m, using Grade 20 concrete and Grade 400 steel.

Chapter 11

Shear and Torsion

All structures are acted on by shear forces as well as bending moments, and in some structures the loads also produce torsion. The overall dimensions obtained from bending moments are generally larger than those obtained from shear or torsion. Only in flat plates and footings are shear forces likely to determine sizes.

However, shear and torsion frequently require additional reinforcement, and failures have occurred because this has been overlooked.

11.1 DETERMINATION OF SHEAR FORCE

We discussed in Chapter 7 the determination of bending moments and noted that they are calculated by an elastic analysis, and that only the stress distribution at the cross section of maximum moment is determined by ultimate strength. The same applies to shear.

If the elastic analysis is undertaken with the aid of a digital computer, shear forces are normally obtained as part of the printout. When moment distribution or bending moment coefficients are used, the shear forces have to be determined separately. In that case the empirical rule in Section 8.3 of the *ACI Code* can be used, which states that the shear force may be taken as

$$\frac{1.15w_u\ell_n}{2} \quad \text{at the first interior support of a multiple span}$$

$$\frac{1.0w_u\ell_n}{2} \quad \text{for any other support}$$

where w_u is the uniformly distributed ultimate load per unit length of beam, or per unit area of slab, and ℓ_n is the length of the clear span, measured face to face of supports.

11.2 SHEAR STRESSES IN CONCRETE BEAMS

The shear force is the resultant of all the vertical forces acting on a beam at the section under consideration (Figure 11.1), and it tends to cut or shear the beam.

This force sets up both horizontal and vertical stresses (Figure 11.2). The horizontal stresses are easily demonstrated if we take a deep timber beam and cut it into a series of planks (Fig. 11.3). The individual planks slide over one another, and the strength of the planks is much less than the beam from which they were cut. If we now grip the planks with large G-clamps so that the sliding action is prevented, the original strength of the beam is restored.

The vertical shear stresses v_1 and the horizontal shear stresses v_2 are equal. We can prove this as follows: Considering the cube measuring $a \times a \times a$ in Fig. 11.2, the vertical shear stresses v_1 act on an area $a \times a$, and thus form a moment v_1a^2a. Similarly, the horizontal shear stresses form a moment v_2a^2a. One of these moments is clockwise and the other is anticlockwise, and they must be exactly equal; otherwise, the element would rotate (which it does not). Consequently,

$$v_1 = v_2 = v_{12}$$

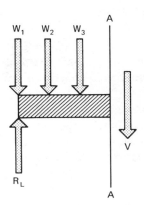

Figure 11.1 The shear force is the resultant of all the vertical forces acting on a beam at the section under consideration, that is, the force the section must provide for vertical equilibrium: $V = R_L - W_1 - W_2 - W_3$.

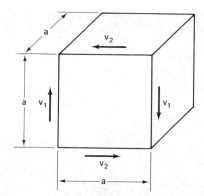

Figure 11.2 A shear force sets up both vertical shear stresses v_1 and horizontal shear stresses v_2.

Figure 11.3 Adjacent horizontal layers of a beam have a tendency to slide over one another, which is resisted by the horizontal shear stresses.

Let us now cut the cube in half (Fig. 11.4) along one diagonal. Evidently, the diagonal components of the horizontal and vertical shear stresses combine to produce a diagonal compressive stress across one diagonal, and a diagonal tensile stress across the other diagonal at right angles. The concrete can easily withstand the diagonal compression, but when the diagonal tensile stress becomes excessive, it produces a diagonal tensile crack (Figure

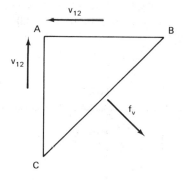

Figure 11.4 The horizontal and vertical shear stresses combine to produce a diagonal tensile stress f_v. Taking a triangle of unit length, the horizontal and vertical shear forces are $v_{12} \times 1^2$, and the diagonal tensile force, acting on the diagonal is $f_v \times 1 \times \sqrt{2}$. Resolving diagonally yields

$$v_{12} \cos 45° + v_{12} \sin 45° = f_v \sqrt{2}$$

$$\frac{v_{12}}{\sqrt{2}} + \frac{v_{12}}{\sqrt{2}} = f_v \sqrt{2}$$

which gives $f_v = v_{12}$.

Figure 11.5 Diagonal tensile failure of reinforced concrete beam due to shear.
The beam was simply supported and carried a single central load.

11.5). Evidently, the shear strength of a reinforced concrete beam is de-
pendent on the tensile strength of the concrete, which by experiment is proportional to $\sqrt{f_c}$, that is, the ratio of tensile to compressive strength is much
less favorable for high-strength than for low-strength concrete.

Since shear failure is actually a diagonal tension failure, the *critical cross
section* for shear in monolithic construction is not at the support itself, unless
the support is such that a diagonal tension crack can extend into it. At a
column junction the diagonal tension is usually suppressed by the compressive
column stresses. A crack of 45° cannot form until we reach a section at a
distance d from the support (Figure 11.6). The critical section for shear is
therefore at a distance d from the support, and the maximum area of shear
reinforcement is calculated for the shear force at that section.

The theory of shear stress distribution, which has been reproduced in
books on reinforced concrete design for more than 50 years (Reference 11.1),
was originated by Emil Mörsch in 1903 (Reference 11.2). It may be correct
until the concrete forms diagonal tension cracks; however, in view of the low

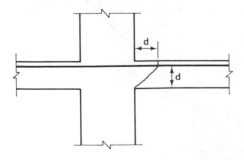

Figure 11.6 The critical cross section
for shear is at a distance d from the
support. Shear failure is a failure of
the concrete in diagonal tension, and a
crack at 45° cannot form until we reach
a section at a distance d from the
support (where d is the effective depth).
This assumes that the diagonal tension
is suppressed at the support (e.g., by
compressive column stresses).

tensile strength of concrete, this covers only a small proportion of practical concrete structures, and the theory of the shear strength of concrete became more and more complicated as research indicated new factors that needed to be considered. A study of Eivind Hognestad in 1952 (Reference 11.3) initiated a more empirical approach, and the *ACI Code* of 1971 (Reference 11.4) replaced the theoretically based formulas by empirical rules. This was followed by the 1977 and 1983 ACI codes.

We first calculate the nominal ultimate shear force

$$V_n = \frac{V_u}{\phi} \tag{11.1}$$

where V_u is the ultimate, load-factored shear force and $\phi = 0.85$ is the strength-reduction factor for shear (see Section 8.6).

The nominal ultimate shear stress is determined by dividing V_n by the width b and the effective depth d:

$$v_n = \frac{V_n}{bd} \tag{11.2}$$

It is not suggested that this is the actual shear stress. This nominal shear stress must not exceed an empirically determined stress which is proportional to the diagonal tensile strength of the concrete.

The maximum shear stress occurs between the neutral axis and the reinforcement (this is proved in most textbooks on structural mechanics, for example, Reference 7.2). Therefore, the width to be considered for the design of shear reinforcement is the width at the neutral axis. Consequently, for flanged beams (whether T-beams or not, see Section 10.3) the nominal ultimate shear stress

$$v_n = \frac{V_n}{b_w d} \tag{11.3}$$

where b_w is the width of the web.

When the ultimate nominal shear stress v_n reaches a critical stress $2\sqrt{f'_c}$ concrete cracks diagonally (Fig. 11.7). Provided that the reinforcement is securely anchored (see Figs. 6.5 and 6.6), the section is kept in equilibrium by the resistance moment formed by the tensile force in the reinforcement T and the compressive force in the undamaged concrete C. However, when

Figure 11.7 Formation of diagonal tension crack in a reinforced concrete beam. The section is kept in equilibrium as long as the reinforcement is securely anchored and the concrete does not crush in compression, but the crack needs to be bridged by shear reinforcement.

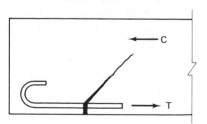

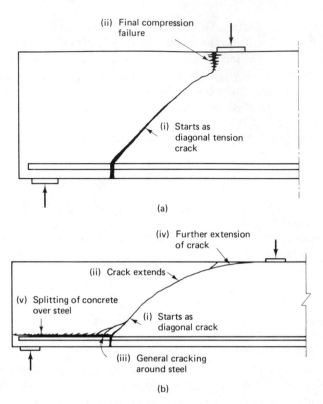

Figure 11.8 Shear failure of reinforced concrete beam. (a) Shear-compression failure occurs by crushing of the concrete in the compression zone. (b) Shear-splitting failure occurs by extension of the diagonal tension crack upward, or splitting of concrete along a horizontal line over the tension steel, or both.

v_n exceeds the tensile strength of the concrete, which is taken in the *ACI Code* as $2\sqrt{f_c'}$, it becomes necessary to provide shear reinforcement (see Section 11.4) for the shear stress in excess of that which could be resisted by the concrete:

$$v_s = v_n\ 2\sqrt{f_c'} \tag{11.4}$$

Even shear reinforcement cannot increase the shear strength of beams indefinitely, because the concrete eventually disintegrates, irrespective of the amount of reinforcement provided. There are two types of ultimate shear failure (Fig. 11.8).

A shear-compression failure occurs by crushing of concrete after the diagonal crack has reduced the compression zone to the extent where it is no longer able to sustain the flexural compressive force C [Figs. 11.7 and 11.8(a)].

A shear-splitting failure [Fig. 11.8(b)] occurs by extension of the diag-

TABLE 11.1 Limiting Shear Stresses (in psi)

	Specified Concrete Strength			
Limit	3000	4000	5000	6000
Shear reinforcement required, $2\sqrt{f'_c}$	110	126	141	155
Closer spacing for shear reinforce-ment, $4\sqrt{f'_c}$	219	253	283	310
Upper limit to shear reinforcement, $8\sqrt{f'_c}$	438	506	566	620

TABLE 11.2 Limiting Shear Stresses (in MPa)

	Specified Concrete Strength			
Limited	20	·25	30	35
Shear reinforcement required, $\frac{1}{6}\sqrt{f'_c}$	0.75	0.83	0.91	1.00
Closer spacing for shear reinforce-ment, $\frac{1}{3}\sqrt{f'_c}$	1.50	1.67	1.83	2.00
Upper limit to shear reinforcement, $\frac{2}{3}\sqrt{f'_c}$	3.00	3.33	3.65	4.00

onal crack upward, which either splits the beam into two separate parts, or more commonly produces splitting of the concrete over the longitudinal steel causing collapse of the beam.   The mechanics of this type of failure is very complex, and attempts at a theoretical analysis have been abandoned.   The *ACI Code* specifies an upper limit $8\sqrt{f'_c}$, and v_s must be less irrespective of the amount of shear reinforcement.   This provision sets an upper limit to the amount of useful shear reinforcement.

There is one further limitation.   When the shear stress v_s reaches $4\sqrt{f'_c}$, the increasing cracking requires a closer spacing of the shear reinforcement (see Section 11.3).   These limiting shear stresses are listed in Table 11.1 (Table 11.2 gives the limits in metric units).

11.3 DETAILS OF SHEAR REINFORCEMENT

Shear reinforcement is mainly of two types (Fig. 11.9):

1. We bend up the longitudinal bars at 45° so that they are at right angles to the direction of diagonal tensile stresses set up by the shear forces. We do not, in that case, provide additional reinforcement for shear, but

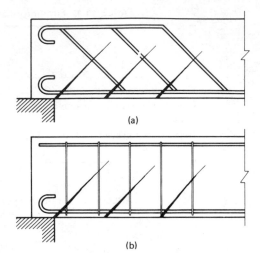

Figure 11.9 Shear reinforcement. (a)
Bars bent up at 45° acting as shear
reinforcement. (b) Vertical stirrups as
shear reinforcement.

utilize the existing reinforcement previously designed for bending. Since shear reinforcement is required over a range, we bend the bars up gradually, usually in pairs. Since the greater part of the main reinforcement is required at the bottom at midspan and at the top over the supports, we may in any case bend up some of the bars.

2. We use ties or stirrups at right angles to the main reinforcement, whose diagonal component resists the diagonal tensile stress. These stirrups can also serve to restrain the buckling of the compression reinforcement, if any (see Example 11.4), and they are useful for forming the reinforcing bars into a cage that facilitates the casting of the concrete.

Vertical shear reinforcement in edge beams and spandrel beams must be in the form of closed stirrups, because such beams are inevitably subject to some torsion, and torsional shear reinforcement must be in the form of closed stirrups (Section 11.7).

Shear stresses are not critical for slabs supported on beams (see Examples 11.1 and 11.2). In footings and flat plates v_n may exceed $2\sqrt{f_c'}$, but it is then usually simpler and cheaper to increase the thickness of the slab or the circumference of the column rather than to provide shear reinforcement (see Sections 12.5 and 14.2).

Shear stresses are also not critical in concrete-joist or hollow-block construction (see Sections 12.5 and 14.2).

In beams a nominal *minimum of shear reinforcement* (Section 11.5.5 of the *ACI Code*) must be provided if the overall depth of the beam is more than 10 in. (250 mm), or $2.5h_f$, or $\frac{1}{2}b_w$ (where h_f is the thickness of the flange and b_w is the width of the web; see Fig. 10.3), unless $v_n < \sqrt{f_c'}$.

This minimum is intended to provide the equivalent of an ultimate shear stress in the concrete of 50 psi ($\frac{1}{3}$ MPa) acting on the width b_w and on the length of beam s (where s is the spacing of the stirrups):

$$50b_w s = f_y A_v$$

where f_y is the yield stress of the steel and A_v is the area of both legs of the stirrup.

The minimum permissible spacing is thus

$$s = \frac{f_y A_v}{50b_w} \tag{11.5a}$$

in traditional British/American units, and

$$s = \frac{3f_y A_v}{b_w} \tag{11.5b}$$

in metric units.

For all shear reinforcement there is a *limitation on the spacing* (Section 11.5.4 of the *ACI Code*) to ensure that each diagonal crack is crossed by at least one shear reinforcing bar.

For vertical shear reinforcement, the maximum permissible spacing is 24 in. (600 mm) or $0.50d$ (whichever is less), where d is the effective depth. For shear reinforcement bent up at 45° the minimum spacing parallel to the axis of the beam is d. These maximum spacings are halved if the shear stress $v_s > 4\sqrt{f_c'}$ (see Table 11.1), but this rarely happens in small building frames.

11.4 DESIGN OF SHEAR REINFORCEMENT

The concrete is capable of resisting a shear force

$$V_c = 2\sqrt{f_c'} b_w d \tag{11.6}$$

where f_c' is the specified compressive strength of the concrete, b_w is the width of the web of the beam or the width of a rectangular beam, and d is the effective depth. The remainder of the nominal ultimate shear force

$$V_s = V_n - V_c \tag{11.7}$$

produce a nominal shear stress

$$v_s = \frac{V_s}{b_w d} \tag{11.8}$$

and that must be resisted by the shear reinforcement.

Let us first consider *bars bent up at 45°* (Fig. 11.10). A_v is the cross-

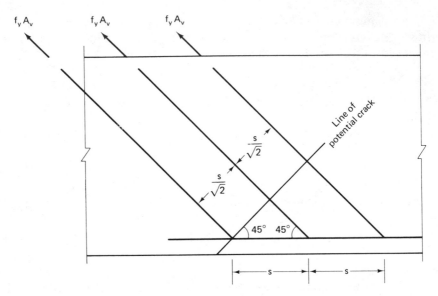

Figure 11.10 Shear reinforcement at right angles to the line of a potential crack (i.e., bars bent up at 45°). The spacing is measured parallel to the axis of the beam.

sectional area of all the bars bent up at the section. The tensile force in these bars stressed to the yield point is

$$f_y A_v$$

This force is required to resist the diagonal tensile stress in concrete across the line of a potential crack for a distance $s/\sqrt{2}$ along the crack, and over a width of beam b_w, so that

$$\frac{v_s b_w s}{\sqrt{2}} = f_y A_v$$

Substituting from Eq. (11.8), the nominal shear force resisted by bars bent up at 45° is

$$V_s = \frac{f_y A_v \sqrt{2} d}{s} \tag{11.9}$$

Let us next consider *vertical stirrups*. A_v is the cross-sectional area of the two vertical legs of a single stirrup, or of four bars for double stirrups. The tensile force in these stirrups stressed to yield stress is $f_y A_v$. The component of this force across the line of the potential crack equals the diagonal tensile force in the concrete:

$$f_y A_v \cos 45° = \frac{f_y A_v}{\sqrt{2}} = \frac{v_s b_w s}{\sqrt{2}}$$

Consequently, the nominal shear force resisted by vertical stirrups is

$$V_s = \frac{f_y A_v d}{s} \qquad (11.10)$$

Normally, there is little choice in bar size, since most stirrups are made from No. 3 bars. Given A_v, the maximum bar spacing is

$$s = \frac{d f_y A_v}{V_s} \qquad (11.11)$$

11.5 DESIGN EXAMPLES FOR SHEAR

It is unnecessary to check reinforced concrete one-way and two-way slabs, because the shear stresses are always low, and shear reinforcement is never required. However, in the following two examples we will check the two slabs designed in Chapter 9 to prove the point.

Flat slabs (Chapter 12) and footing slabs (Chapter 14) need checking, but if the shear stress is too high, it is generally simpler to make the slab deeper rather than use reinforcement.

Beams always need shear reinforcement if they are more than 10 in. (250 mm) deep. The stirrups also serve as supports for the main reinforcement (Figs. 9.8 and 10.13 to 10.15) and as ties for compression reinforcement, if any. Closed stirrups can also resist torsion (see Section 11.6).

Example 11.1A

Check the slab of Example 9.2 for shear.

Data given: $D = 100$ psf, $L = 60$ psf, $\ell_n = 13.5$ ft, $b = 1$ ft, $f'_c = 3500$ psi, $f_y = 60{,}000$ psi, $\phi = 0.85$, $h = 7$ in., and $A_s = $ No. 4 at 9 in.

The ultimate load is

$$w_u = 1.4 \times 100 + 1.7 \times 60 = 242 \text{ psf}$$

The maximum ultimate shear force (Section 11.1) is

$$V_u = \frac{1.15 w_u \ell_n}{2} = \frac{1.15 \times 242 \times 13.5}{2} = 1879 \text{ lb/ft}$$

From Eq. (11.1), the nominal ultimate shear force is

$$V_n = \frac{V_u}{\phi} = \frac{1879}{0.85} = 2210 \text{ lb/ft}$$

The effective depth is

$$d = h - 0.75 - \tfrac{1}{2} \times 0.50 = 6.0 \text{ in.}$$

The nominal ultimate shear stress, from Eq. (11.2), is

$$v_n = \frac{V_n}{bd} = \frac{2210}{12 \times 6} = 30.7 \text{ psi}$$

The maximum admissible ultimate shear stress is $2\sqrt{f'_c} = 118$ psi (Table 11.1), which is almost four times the actual ultimate shear stress.

Example 11.1B

Check the slab of Example 9.1.B for shear.
    Data given: $D = 5$ kPa, $L = 3$ kPa, $\ell_n = 4$ m, $b = 1$ m, $f'_c = 25$ MPa, $f_y = 400$ MPa, $\phi = 0.85$, $h = 175$ mm, and $A_s = $ No. 10 at 175 mm.
    The ultimate load is

$$w_u = 1.4 \times 5 + 1.7 \times 3 = 12.1 \text{ kPa}$$

The maximum ultimate shear force (Section 11.1) is

$$V_u = \frac{1.15 w_u \ell_n}{2} = \frac{1.15 \times 12.1 \times 4}{2} = 27.83 \text{ kN/m}$$

From Eq. (11.1), the nominal ultimate shear force is

$$V_n = \frac{V_u}{\phi} = \frac{27.83}{0.85} = 32.74 \text{ kN/m} = 0.033 \text{ MN/m}$$

The effective depth is

$$d = h - 20 - \tfrac{1}{2} \times 10 = 150 \text{ mm}$$

The nominal ultimate shear stress, from Eq. (11.2), is

$$v_n = \frac{V_n}{bd} = \frac{0.033}{1.0 \times 0.150} = 0.22 \text{ MPa}$$

The maximum admissible ultimate shear stress is $\tfrac{1}{6}\sqrt{f'_c} = 0.83$ MPa from Table 11.2, which is almost four times the actual ultimate shear stress.

Example 11.2

Check the slab of Example 9.4.A for shear.
    Data given: $D = 110$ psf, $L = 60$ psf, $\ell_n = 13.5$ ft, $b = 1$ ft, $f'_c = 3500$ psi, $f_y = 60,000$ psi, $\phi = 0.85$, $h = 8.5$ in., and $A_s = $ No. 4 bars at 12 in.
    The ultimate load is

$$w_u = 1.4 \times 110 + 1.7 \times 60 = 256 \text{ psf}$$

The maximum ultimate shear force (Section 11.1) is

$$V_u = \frac{1.0 \times w_u \ell_n}{2} = \frac{256 \times 13.5}{2} = 1728 \text{ lb/ft}$$

The nominal shear force is

$$V_n = \frac{1728}{0.85} = 2033 \text{ lb/ft}$$

The effective depth is

$$d = 8.5 - 0.75 - \tfrac{1}{2} \times 0.50 = 7.50 \text{ in.}$$

The nominal ultimate shear stress is

$$v_n = \frac{2033}{12 \times 7.5} = 22.6 \text{ psi (118 psi permitted)}$$

Example 11.3A

Check the beam of Example 9.6.A for shear.

    Data given: D = 1000 lb/ft, L = 500 lb/ft, ℓ = 30 ft, f'_c = 4000 psi, f_y = 60,000 psi, ϕ = 0.85, b = 12 in., h = 24 in., and A_s = 4 No. 8 bars.

    The ultimate load is

$$w_u = 1.4 \times 1000 + 1.7 \times 500 = 2250 \text{ lb/ft}$$

The critical section for shear (Fig. 11.6) is at a distance d from the face of the support. As an approximation, which errs on the safe side, we will determine the maximum ultimate shear force at the support, since the actual stirrup spacing is unlikely to be determined by the magnitude of the shear force:

$$V_u = \frac{w_u \ell}{2} = \frac{2250 \times 30}{2} = 33,750 \text{ lb}$$

The nominal ultimate shear force is

$$V_n = \frac{33,750}{0.85} = 39,700 \text{ lb}$$

The effective depth is

$$d = h - 1.5 \text{ (cover)} - 0.38 \text{ (stirrup diameter)} - \tfrac{1}{2} \times 1.0 \text{ (bar)} = 21.62 \text{ in.}$$

$$v_n = \frac{39,700}{12 \times 21.62} = 153 \text{ psi}$$

From Table 11.1, the concrete can resist 126 psi, and the balance must be resisted by shear reinforcement. This excess shear stress is

$$v_s = 153 - 126 = 27 \text{ psi}$$

Nominal shear reinforcement must be provided throughout beams more than 10 in. deep to produce the equivalent of an ultimate shear stress of 50 psi. This is more than the reinforcement required by the maximum shear force.

    Using single No. 3 stirrups,

$$A_v = 0.22 \text{ in.}^2 \quad \text{(2 bars)}$$

and the maximum bar spacing required by Eq. (11.5) is

$$s = \frac{f_y A_v}{50b} = \frac{60,000 \times 0.22}{50 \times 12} = 22 \text{ in.}$$

But the *maximum* permissible bar spacing is 24 in. or $\frac{1}{2}d$, whichever is less; that is

$$s = \frac{1}{2} \times 21.62 = 10.81 \text{ in.}$$

Bars at 10-in. centers provide more shear reinforcement than the same bars at 11-in. centers. *Use No. 3 stirrups at 10-in. centers.*

If the beam is at the edge of a building, torsion must be considered, and closed stirrups must be used (see Section 11.6 and Example 11.5).

Example 11.3B

Check the beam of Example 9.6B for shear.

Data given: $D = 12$ kN/m, $L = 7$ kN/m, $\ell = 10$ m, $f'_c = 25$ MPa, $f_y = 400$ MPa, $\phi = 0.85$, $b = 300$ mm, $h = 650$ mm, and $A_s = 4$ No. 25 bars.

The ultimate load is

$$w_u = 1.4 \times 12 + 1.7 \times 7 = 28.7 \text{ kN/m}$$

The critical section for shear (Fig. 11.6) is at a distance d from the face of the support. As an approximation, which errs on the safe side, we will determine the maximum ultimate shear force at the support, since the actual stirrups spacing is unlikely to be determined by the magnitude of the shear force:

$$V_u = \frac{w_u \ell}{2} = \frac{28.7 \times 10}{2} = 143.5 \text{ kN}$$

The nominal ultimate shear force is

$$V_n = \frac{143.5}{0.85} = 168.82 \text{ kN}$$

The effective depth is

$$d = h - 40 \text{ (cover)} - 10 \text{ (stirrup diameter)} - \frac{1}{2} \times 25 \text{ (bar)} = 587 \text{ mm}$$

$$v_n = \frac{168.82 \times 10^{-3}}{0.3 \times 0.587} = 0.959 \text{ MPa}$$

From Table 11.2, the concrete can resist 0.83 MPa, and the balance must be resisted by shear reinforcement. This excess shear stress is

$$v_s = 0.959 - 0.83 = 0.129 \text{ MPa}$$

Nominal shear reinforcement must be provided throughout beams more than 250 mm deep to produce the equivalent of an ultimate shear stress of $\frac{1}{3}$ MPa. This is more than the reinforcement required by the maximum shear force.

Using single No. 10 stirrups,

$$A_v = 200 \text{ mm}^2 \text{ (2 bars)}$$

and the maximum bar spacing required by Eq. (11.5) is

$$s = \frac{3 f_y A_v}{b} = \frac{3 \times 400 \times 200}{300} = 800 \text{ mm}$$

But the *maximum* permissible bar spacing is 600 mm or $\frac{1}{2}d$, whichever is less; that is,

$$s = \frac{1}{2} \times 587 = 293.5 \text{ mm}$$

Bars at 275-mm centers provide more shear reinforcement than the same bars at 300-mm centers. *Use No. 10 stirrups on 275-mm centers.*

If the beam is at the edge of a building, torsion must be considered, and closed stirrups must be used (see Section 11.6).

Example 11.4

Check the beam of Example 9.9.A for shear.

Data given: $D = 1000$ lb/ft, $L = 500$ lb/ft, $\ell = 30$ ft, $f_c' = 4000$ psi, $f_y = 60,000$ psi, $\phi = 0.85$, $b = 12$ in., $h = 24$ in., and $A_s = 2$ No. 9 + 2 No. 6 bars maximum.

The ultimate load is

$$w_u = 1.4 \times 1000 + 1.7 \times 500 = 2250 \text{ lb/ft}$$

In a continuous beam the critical shear force occurs at a distance d from the support (Fig. 11.6), and at the first interior support the shear force must be multiplied by 1.15 (Section 11.1). However, it is likely that only nominal stirrups will be needed, and that a precise calculation is superfluous. Subject to later revision, we will therefore take the shear force to be

$$V_u = \frac{1}{2}w_u\ell = \frac{1}{2} \times 2250 \times 30 = 33,750 \text{ lb}$$

The nominal ultimate shear force is

$$V_n = \frac{33,750}{0.85} = 39,700 \text{ lb}$$

The effective depth is

$$d = 24 - 1.5 - 0.38 - \frac{1}{2} \times 1.13 = 21.55 \text{ in.}$$

$$v_n = \frac{39,700}{12 \times 21.55} = 154 \text{ psi}$$

The concrete can resist 126 psi (Table 11.1) and 50 psi worth of nominal shear reinforcement is additionally required throughout the beam. Only nominal shear reinforcement is therefore needed. The smallest bar size is No. 3, whose $A_v = 0.22$ in.^{2} (for 2 legs of a single stirrup), and the maximum permissible bar spacing is 24 in. or $\frac{1}{2}d = 9.78$ in., say 9 in. This gives a nominal shear stress, from Eqs. (11.8) and (11.10), of

$$\frac{f_y A_v}{bs} = \frac{60,000 \times 0.22}{12 \times 9} = 122 \text{ psi}$$

whereas only 50 psi is required. *Use No. 3 stirrups on 9-in. centers.*

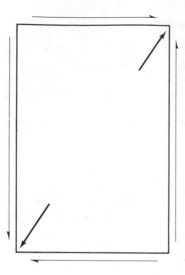

Figure 11.11 Action of shear wall. Since the horizontal forces due to wind or earthquakes may act in either direction, the reinforcement must be designed to function in both directions. Thus vertical and horizontal reinforcing bars, or a welded mesh, are more appropriate than diagonal bars.

11.6 SHEAR WALLS

Walls (see Section 13.11) can be utilized for resisting horizontal forces due to wind and earthquakes. The moment due to the horizontal forces (Fig. 11.11) is resisted by an equal and opposite moment due to vertical shear forces (see Fig. 11.2), and these combine to produce diagonal tension in the wall. This must be resisted by reinforcement. The *ACI Code* in the Appendix A on Seismic Design, Section A.8, recommends a minimum vertical and horizontal reinforcement ratio

$$\rho = 0.0025$$

calculated on the gross area in each direction, and a minimum area of vertical reinforcement concentrated near the ends of the wall, so that

$$\rho = \frac{200}{f_y}$$

11.7 TORSION IN CONCRETE STRUCTURES

Torsion is very common in concrete structures, but it is generally of secondary importance. Small twisting moments occur in edge beams because they are loaded on one side only. Hence for edge beams and spandrels vertical shear reinforcement must be in the form of closed stirrups.

Larger twisting moments occur when a beam carries an eccentric curtain wall or facade (Fig. 11.12). Very large twisting moments occur with unsymmetrically arranged beams (Figs. 11.13 and 11.14), in balcony girders (Fig. 11.15), and in spiral staircases. The analysis of structural systems for torsion

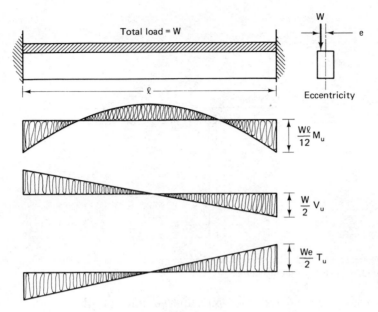

Figure 11.12   Variation of bending moment, transverse shear force, and twisting moment in a spandrel beam restrained at the ends and supporting a brick wall lined up with the edge of the spandrel, and placed eccentrically on it.

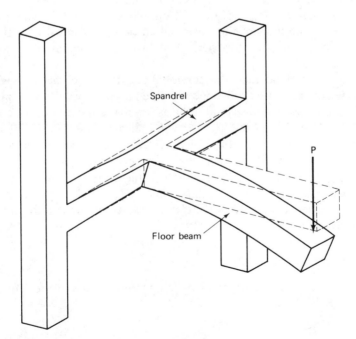

Figure 11.13   A beam framing into a spandrel between columns produces torsion in the spandrel.

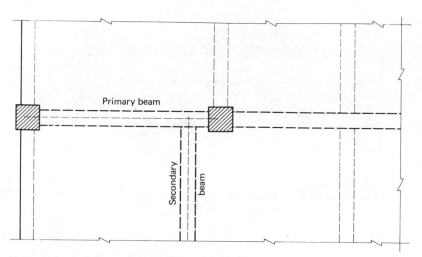

Figure 11.14 Secondary beams framing into primary beams off the column can produce high twisting moments.

has been discussed in detail in a symposium volume published by the American Concrete Institute (Ref. 11.5).

When a plain concrete beam is twisted, it fails along a helical line at 45° to the axis when the diagonal tensile stress exceeds the tensile strength of the concrete (Fig. 11.16). Torsion is thus a form of shear.

Failure can be delayed, but not entirely prevented, by shear reinforcement crossing the lines of the potential cracks. This must take the form of closed stirrups, since the torsional shear stresses occur on all four faces of a rectangular beam.

On one vertical face the torsional shear stresses act opposite to the transverse shear stresses which we discussed in Section 11.2; but on the other vertical face they are additive, and this is where the failure occurs. The upper limit of the shear stress due to transverse shear alone, due to torsion alone, and due to the combined effect of transverse shear and torsion is the same. Torsion consequently decreases the shear capacity of a beam, and vice versa.

Design for torsion is very simple if the combined shear stress due to torsion and transverse shear is less than $2\sqrt{f_c'}$ (see Section 11.2). This is most easily checked by converting the twisting moment into an equivalent shear force:

$$V_t = \frac{1.6T_n}{b} \tag{11.12}$$

where $T_n = T_u/\phi$ is the nominal twisting moment due to the ultimate loads.

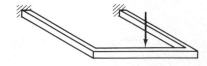

Figure 11.15 Balcony beams are subject to combined bending and torsion.

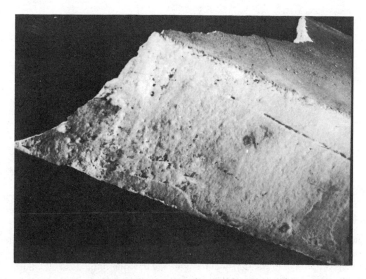

Figure 11.16   Failure of a plain concrete beam in pure torsion.   The diagonal tension crack starts to form at the middle of one of the longer sides, and spirals around the beam on three sides, at an angle of 45°, at right angles to the direction of diagonal tensile stress.   On the fourth side the fracture is of necessity at a different angle.

This equation is obtained by computing the maximum shear stress due to T_n, and determining a shear force V_t which produces the same shear stress (Ref. 11.6).

If we add this to the nominal ultimate shear force V_n, we can calculate the nominal shear stress due to the combined effect of shear and torsion.

$$v_n = \frac{V_n + V_t}{bd} \leq 2\sqrt{f_c'} \tag{11.13}$$

If this combined stress exceeds $2\sqrt{f_c'}$, shear reinforcement is required.

The theoretical background of torsional shear reinforcement has been discussed in detail elsewhere (Refs. 11.7 and 11.8).

It is important that beams subject to torsion should be reinforced with closed stirrups properly anchored by hooks, and not with open stirrups, because the torsional shear stresses, unlike the flexural shear stresses, occur on the horizontal as well as on the vertical faces.   It is also important that there should be a longitudinal bar in *each* corner of the stirrup to take the horizontal component of the diagonal tension due to torsion (Fig. 11.17).   If reinforcement has already been provided to resist a bending moment, this is often sufficient to resist the additional stresses due to the, usually small, twisting moments.   One No. 4 bar in each corner should be sufficient for torsion, unless it is exceptionally high.

The *ACI Code*, in Section 11.5.5.5, requires that the combined area A_v of the closed stirrups for shear and torsion should provide the equivalent of a shear stress 50 psi (⅓ MPa), the same as for shear alone.   The author

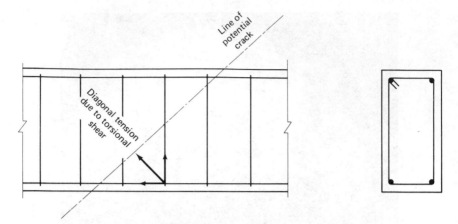

Figure 11.17. Torsional shear reinforcement normally takes the form of closed vertical stirrups, with a longitudinal bar in each corner. The stirrups resist the vertical component of the diagonal tension, and the longitudinal bars the horizontal component. Anchorage is critical for the vertical stirrups.

considers that it would be wise to make a slightly higher provision for spandrel beam, say 60 psi (0.4 MPa). Equation (11.5) then becomes

$$s = \frac{f_y A_v}{60 b_w} \qquad (11.14)$$

If a building carries only the usual loads, and there are no exceptional load eccentricities or load concentrations, further calculations are unnecessary when the calculations for flexural shear show that only the nominal reinforcement is required (Examples 11.3 and 11.4).

Example 11.5

The beam of Examples 9.6 and 11.3 was assumed to be loaded symmetrically about its center line. We will assume that the load is not symmetrical, but has an eccentricity of 1 inch relative to the center line, so that the beam is slightly twisted by the load. Determine the torsional reinforcement required, if any.

Data given: $D = 1000$ lb/ft, $L = 500$ lb/ft, $\ell = 30$ ft, $f'_c = 4000$ psi, $f_y = 60,000$ psi, $\phi = 0.85$, $b = 12$ in., $h = 24$ in., $A_s = 4$ No. 8 bars, and $e = 1$ in.

In Example 11.3.A we calculated that the effective depth $d = 21.62$ in., the ultimate load $w_u = 2250$ lb/ft, and the nominal ultimate shear force $V_n = 39,700$ lb. The twisting moment caused by the eccentricity of 1 in. is

$$T_u = w_u \ell e = 2250 \text{ lb/ft} \times 30 \text{ ft} \times 1 \text{ in.} = 67,500 \text{ lb·in.}$$

The nominal twisting moment is

$$T_n = \frac{67,500}{0.85} = 79,400 \text{ lb·in.}$$

The equivalent shear force is

$$V_t = \frac{1.6T_n}{b} = \frac{1.6 \times 79,400}{12} = 10,588 \text{ lb}$$

The shear stress due to the combined effect of shear and torsion is

$$v_n = \frac{V_n + V_t}{bd} = \frac{39,700 + 10,588}{12 \times 21.62} = 194 \text{ psi}$$

From Table 11.1, the concrete can resist 126 psi, and the rest must be taken by reinforcement. This excess shear stress is

$$v_s = 194 - 126 = 68 \text{ psi}$$

This is more than 50 psi, so more than nominal shear reinforcement is required.
Using single No. 3 stirrups,

$$A_v = 0.22 \text{ in.}^2 \quad \text{(2 bars)}$$

and the maximum spacing required from Eqs. (11.8) and (11.11) is

$$s = \frac{f_y A_v}{v_s b_w} = \frac{60,000 \times 0.22}{68 \times 12} = 16.2 \text{ in.}$$

The maximum permissible spacing of the shear reinforcement is 24 in. or $\frac{1}{2}d = 10.81$, whichever is the smaller. Consequently, the torsion resulting from the eccentricity does not make any difference to the spacing calculated in Example 9.3. *Use No. 3 stirrups at 10-in. centers.*

However, because torsion is present, the stirrups must be in the form of closed hoops (Fig. 11.17), and adequate longitudinal reinforcement must also be provided. We have sufficient longitudinal reinforcement at the bottom to resist the bending moment, but we also need some on top. *Use 2 No. 4 bars, one in each top corner of the closed hoops, as shown in Fig. 11.17.*

EXERCISES

1. Draw a diagram showing that vertical loads acting on (horizontal) beams and slabs produce both bending moments and shear forces.
2. The principal dimensions of reinforced concrete structural members (with a few exceptions such as flat plates and foundation slabs, discussed in subsequent chapters) are determined by the magnitude of the bending moments, not the shear forces. Why is this so?
3. The vertical shear forces set up both vertical and horizontal shear stresses, which in turn can be combined into diagonal tensile and diagonal compressive stresses. Why can the diagonal tensile stresses cause failure of the concrete, whereas the diagonal compressive stresses are relatively harmless? In structural steel plate girders the diagonal tension is relatively harmless, but the diagonal compression is liable to cause web buckling. What is the reason for the different behavior of steel and concrete structures?

4. In designing a beam or slab to resist the bending moment, we assumed (Section 8.5) that the tensile strength of the concrete may be neglected. In designing the same section to resist the shear force, we consider the tensile strength of the concrete. What is the reason for the difference?

5. Describe the mode of failure of a reinforced concrete beam in shear (a) as a shear-compression failure, and (b) as a shear-splitting failure.

6. When the maximum shear force occurs at the junction of a beam with its support, we do not take the critical section for the design of the shear reinforcement *at* the support, but at a distance d from the support, where the shear force is less than its maximum value. Why do we do that?

7. There are three limiting shear stresses: $2\sqrt{f_c'}$, $4\sqrt{f_c'}$, and $8\sqrt{f_c'}$. What is the significance of each of these stresses?

8. Why are reinforced concrete slabs (except flat slabs and flat plates) not normally checked for shear?

9. The *ACI Code* requires a minimum amount of shear reinforcement for beams (but not for slabs). What is this minimum?

10. What are the criteria for the maximum permissible spacing of shear reinforcement?

11. In theory, bars bent up at an angle of 45° make more efficient use of the reinforcing steel than vertical stirrups. Why are stirrups normally more economical?

12. Is it permissible to use the same reinforcement for resisting shear and for restraining the compression reinforcement against buckling?

13. What is the structural function of a shear wall? Does a shear wall necessarily look different from any other concrete wall?

14. Describe some structural layouts that are likely to cause torsion in beams.

15. How does reinforced concrete fail in torsion? What is the difference between a normal shear failure (due to the vertical loads acting at the center of the beam) and a torsional shear failure?

16. Why is it necessary to use closed vertical stirrups when torsional shear is involved and also to provide at least one longitudinal bar in each corner? Why are bent-up bars and open stirrups useless for resisting torsional shear?

17. A reinforced concrete beam carries a dead load of 1200 lb/ft and a live load of 700 lb/ft over a clear span of 26 ft. The beam is 16 in. wide and 28 in. deep (overall), and it is reinforced with one layer of No. 9 bars. Determine the amount of shear reinforcement required if the concrete is Grade 3.5 and the reinforcement is Grade 60.

18. Determine the amount of shear reinforcement required if the live load in Exercise 17 increases to 1 000 lb/ft.

REFERENCES

11.1. D. Peabody, *The Design of Reinforced Concrete Structures*, 2nd ed., Wiley, New York, 1946, pp. 40–60.

11.2. E. Mörsch, "Experiments on shear stresses in reinforced concrete beams," *Beton und Eisen* (Berlin), Vol. 2, No. 4, October 1903, pp. 204–207.

11.3. E. Hognestad, *What Do We Know about Diagonal Tension and Web Reinforcement in Concrete?* University of Illinois Experiment Station, Circular Series, No. 64, Urbana, Ill., 1952.

11.4. *Building Code Requirements for Reinforced Concrete (ACI 318–71)*, pp. 36–38, and *Commentary*, pp. 46–49, American Concrete Institute, Detroit, 1971.

11.5. *Analysis of Structural Systems for Torsion*, Publication SP-35, American Concrete Institute, Detroit, 1973.

11.6. A. S. Hall and F. E. Archer, "The SAA Rules for Torsion in Reinforced Concrete Beams," *Civil Engineering Transactions of the Institution of Engineers, Australia*, Vol. CE17, No. 1, 1975, pp. 1–6.

11.7. H. J. Cowan, *Reinforced and Prestressed Concrete in Torsion*, St. Martins Press, New York, 1965.

11.8. *Torsion of Structural Concrete*, Publication SP-18, American Concrete Institute, Detroit, 1968.

Chapter 12

Reinforced Concrete Slabs Spanning in Two Directions

We considered the design of one-way slabs in Chapter 9, and that of their supporting beams in Chapter 10. In Section 12.2 we discuss the design of two-way slabs; the design of their supporting beams is the same as for one-way slabs.

This chapter also deals with ribbed slabs, the use of hollow blocks in slabs, and the design of flat slabs and flat plates that delete the supporting beams.

12.1 ONE-WAY SLABS

Both one-way and two-way slabs are supported on beams on all four edges. The proportion of the loads carried by the supporting beams is roughly as shown in Fig. 12.1.

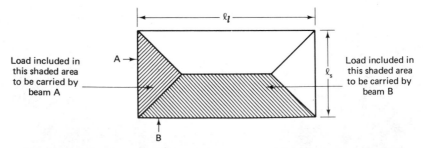

Figure 12.1 Approximate load carried by the supporting beams in a two-way slab.

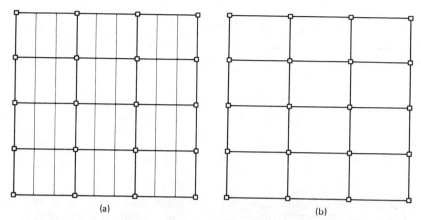

Figure 12.2   One-way and two-way slabs.   (a) One-way slabs spanning across secondary beams which are carried on primary beams.   (b) Two-way slabs supported directly on primary beams.

When $\ell_l/\ell_s > 2$ the contribution made by the span ℓ_l becomes negligible, and we consider that the slab spans across ℓ_s.   We have considered the design of one-way slabs in Chapter 9.

One-way slabs result when a column panel is subdivided by secondary beams.   The secondary beams then transmit the load to the primary beams, which in turn transmit it to the columns [Fig. 12.2(a)].   Subdivision is usually into three spans.   Subdivision into four panels would make the ratio of the primary to the secondary span too great.   Subdivision into two panels would place the only secondary beams to be carried by the primary beams at midspan, where they cause an uneconomically high bending moment.

It is necessary to provide secondary reinforcement both to distribute the loads and to resist shrinkage stresses.

12.2 TWO-WAY SLABS

When the ratio of the longer to the shorter spans drops below 2, and particularly if it drops below 1.5, it becomes worthwhile to consider the contribution made by the longer span.   Two-way slabs are thus supported entirely on primary beams [Fig. 12.2(b)].

The slab can be regarded as being divided into unit strips spanning in both directions, so that each square element of the slab forms part of two strips at right angles to one another (Fig. 12.3).   Let us select the element at the center of the slab.   It forms part of a strip spanning in the s-direction and of another strip spanning in the ℓ-direction, and the central deflection is thus the same for both strips (see Section 8.7).

$$\frac{5}{384} \frac{w_l \ell_l^4}{EI} = \frac{5}{384} \frac{w_s \ell_s^4}{EI}$$

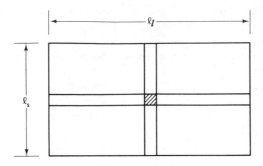

Figure 12.3   Theory of simply supported two-way slabs. The central element shown shaded forms part of one strip spanning across ℓ_l and another spanning across ℓ_s.

Consequently, $w_l\ell_l^4 = w_s\ell_s^4$, and also the total load $w = w_l + w_s$. This means that the proportion of the total load w carried on the span ℓ_s is

$$w_s = \frac{w\ell_l^4}{\ell_s^4 + \ell_l^4}$$

and the bending moment in the span ℓ_s is

$$M_s = \frac{w_s\ell_s^2}{8} = \alpha_s w\ell_s^2$$

and the bending moment in the long span ℓ_l is

$$M_l = \frac{w_l\ell_l^2}{8} = \beta_l w\ell_l^2 = \alpha_l w\ell_s^2$$

If the two-way slab is continuous, which is the more common case, the theory becomes more complicated, but the problem is basically similar.

The resulting bending moment coefficients are shown in Table 12.1. They were published in the 1963 *ACI Code*, but have been omitted from recent ACI codes. The use of these bending moment coefficients is much simpler, if less accurate, than the "equivalent frame method," which considers the slab system as a series of frames interlocking at right angles in the long (*l*) and short (*s*) directions.

The reinforcement in two-way slabs is arranged in middle strips, which comprise the center half of the slab, and column strips of the same width, comprising the other half, that is, the quarter slab on either side of the column. At the edge of the structure the column strip is only half as wide.

The bending moment in the middle strips is

$$M = \alpha w\ell_s^2 \tag{12.1}$$

where α is the coefficient listed in Table 12.1, w the load per unit area carried by the slab, and ℓ_s the span in the shorter direction, or span in either direction for a square slab. This applies to the bending moments calculated for both the long and the short directions.

TABLE 12.1 Bending Moment Coefficients for the Design of Two-way Slabs by the Direct Method[a]

	Moment Coefficient α for Short Span ℓ_s for Ratio of Short Span to Long Span, ℓ_s/ℓ_l						Moment Coefficient α for long Span, ℓ_l, All Span Ratios
	1.0	0.9	0.8	0.7	0.6	0.5 and less	
Case 1: Interior panels							
Negative moment at:							
Continuous edge	0.033	0.040	0.048	0.055	0.063	0.083	0.033
Discontinuous edge	—	—	—	—	—	—	—
Positive moment at midspan	0.025	0.030	0.036	0.041	0.047	0.062	0.025
Case 2: One edge discontinuous							
Negative moment at:							
Continuous edge	0.041	0.048	0.055	0.062	0.069	0.085	0.041
Discontinuous edge	0.021	0.024	0.027	0.031	0.035	0.042	0.021
Positive moment at midspan	0.031	0.036	0.041	0.047	0.052	0.064	0.031
Case 3: Two edges discontinuous							
Negative moment at:							
Continuous edge	0.049	0.057	0.064	0.071	0.078	0.090	0.049
Discontinuous edge	0.025	0.028	0.032	0.036	0.039	0.045	0.025
Positive moment at midspan	0.037	0.043	0.048	0.054	0.059	0.068	0.037
Case 4: Three edges discontinuous							
Negative moment at:							
Continuous edge	0.058	0.066	0.074	0.082	0.090	0.098	0.058
Discontinuous edge	0.029	0.033	0.037	0.041	0.045	0.049	0.029
Positive moment at midspan	0.044	0.050	0.056	0.062	0.068	0.074	0.044
Case 5: Four edges discontinuous							
Negative moment at:							
Continuous edge	—	—	—	—	—	—	—
Discontinuous edge	0.033	0.038	0.043	0.047	0.053	0.055	0.033
Positive moment at midspan	0.050	0.057	0.064	0.072	0.080	0.083	0.050

Source: Reproduced from the 1963 *ACI Code.*
[a]Bending moment $= \alpha w \ell_s^2$ for both short and long spans.

The bending moment in the column strips is

$$M = \tfrac{2}{3} \alpha w \ell_s^2 \qquad (12.2)$$

The minimum thickness requirements of Section 9.5.3 of the *ACI Code*, already cited in this book in Section 8.7, must be satisfied to limit deflection. If that thickness is greater than the thickness required by the maximum bending moment, the reinforcement is often determined by the minimum requirement of Section 7.12 of the *ACI Code*, cited in Section 6.3, which is $0.0018bh$ for Grade 60 steel (0.002 for Grade 40 steel).

There is no secondary reinforcement in two-way slabs, only two sets of primary reinforcement at right angles to one another, that serve the bending moment in the two directions, and also provide the reinforcement to resist temperature and shrinkage stresses.

Two-way slabs can also be designed by the yield-line method (Section 7.5).

12.3 TWO-WAY SLAB EXAMPLE

Example 12.1A

Design the corner panel of a continuous floor structure consisting of beams and two-way slabs, spanning 20 ft between beam centers in both directions. The dead load is 100 psf, the live load is 60 psf, the concrete is Grade 4, and the steel is Grade 60.
    Data given: $D = 100$ psf, $L = 60$ psf, $\ell = 20$ ft, $b = 1$ ft, $f'_c = 4000$ psi, $f_y = 60,000$ psi, and $\phi = 0.90$.
    The ultimate load is

$$w_u = 1.4 \times 100 + 1.7 \times 60 = 242 \text{ psf}$$

Let us assume that the supporting beams are 12 in. (1 ft) wide, so that the clear span is

$$\ell_n = 19 \text{ ft} = 228 \text{ in.}$$

Minimum depth: Control of deflection limits the overall depth (see Section 8.7). From Table 8.3, $h \geqslant 3\tfrac{1}{2}$ in. From Eq. (8.5),*

$$h \geqslant \frac{\ell_n}{36,000}\left(800 + \frac{f_y}{200}\right) = \frac{228 \times 1100}{36,000} = 6.97 \text{ in.}$$

Use a 7 in. slab.
    Minimum reinforcement: The minimum amount of reinforcement to resist temperature and shrinkage stresses (Section 6.3) is

$$A_s = 0.0018bh = 0.0018 \times 12 \times 7 = 0.151 \text{ in.}^2/\text{ft}$$

This requires No. 4 bars on no more than 15-in. centers (0.16 in.²/ft).

---

*This is Eq. (9.13) of the *ACI Code*; h could be reduced to 6.12 in. (giving a $6\tfrac{1}{2}$-in. slab) by using the more complicated Eq. (9.11) of the *ACI Code*.

Reinforcement required by bending moment: The reinforcement in one direction must lie above the reinforcement in the other direction.   Assuming No. 4 bars both ways, the lesser effective depth is

$d = 7 - 0.75$ (cover) $- 0.5$ (outer bars) $- \frac{1}{2} \times 0.5$ (inner bars) $= 5.50$ in.

Let us assume conservatively (Table 9.1) that

$$z = 0.90d = 4.95 \text{ in.}$$

Using the direct method, the maximum bending moments in the middle strip, from Table 12.1, are

$$\text{Negative moment at continuous edge} = 0.049 w_u \ell_s^2 = 0.049 \times 228 \times 20^2$$

$$= 4469 \text{ lb} \cdot \text{ft/ft} = 53{,}625 \text{ lb} \cdot \text{in./ft}$$

$$\text{Negative moment at discontinuous edge} = 0.025 w_u \ell_s^2 = 27{,}360 \text{ lb} \cdot \text{in./ft}$$

$$\text{Positive moment at midspan} = 0.037 w_u \ell_s^2 = 40{,}493 \text{ lb} \cdot \text{in./ft}$$

The maximum bending moments in the column strip are two-thirds of these values.
The absolute maximum bending moment requires, from Eq. (9.8), a steel area of

$$A_s = \frac{M_u}{\phi f_y z} = \frac{53{,}625}{0.9 \times 60{,}000 \times 4.95} = 0.200 \text{ in.}^2/\text{ft}$$

Use No. 4 bars at 12-in. centers at continuous edge ($= 0.20$ in.^{2}/ft).
Selection of reinforcement: Construction would become complicated if we used in the middle strip in conjunction with the 12-in. spacing any spacing other than 12 in. or 24 in. for the other bending moments.   But the minimum area of reinforcement calculated above is No. 4 at 15 in. Therefore, it is appropriate to use the 12-in. spacing also for the lower bending moments throughout the middle strip.   *Use No. 4 bars at 12-in. centers throughout the middle strip.*
In the column strip bending moments are reduced by one-third, and we require a steel area of

$$A_s = \frac{2}{3} \times 0.200 = 0.133 \text{ in.}^2/\text{ft}$$

Use No. 4 bars at 15-in. centers throughout the column strips ($= 0.16$ in.^{2}/ft).
The middle strip is $\frac{1}{2} \times 228 = 114$ in. wide, and thus requires

$$\frac{114}{12} = 9.9 \quad \text{(say 10) reinforcing bars for the bending moment in each direction}$$

The column strips are each 57 in. wide and require

$$\frac{57}{15} = 3.8 \quad \text{(say 4) reinforcing bars each}$$

Arrangement of reinforcement: The bars can either be run straight at the top for the negative bending moments, and straight at the bottom for the positive bending moment, or we can use bars bent up alternatively left and right, each type of bar at a 24 in. spacing in the middle strip and at 30 in. spacing in the column strips.   The second arrangement requires expensive bar bending, but it uses less steel, and makes

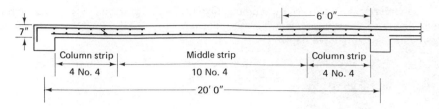

Figure 12.4   Cross section of the slab designed in Example 12.1.A.

it easier to fix the negative reinforcement. This more complicated arrangement is shown in Fig. 12.4.

We will take half the positive reinforcement to the face of the support, and bend up the other half at the point of inflection. A glance at Table 6.1 shows that we have adequate development length without the need of a numerical check. Half the negative reinforcement is bent down, and the other half taken a distance ℓ_a past the point of inflection. We will assume, conservatively for negative reinforcement, that the point of inflection is

$$\tfrac{1}{4}\,\ell_n = \tfrac{1}{4} \times 228 = 57 \text{ in.}$$

from the support. The additional length is

$$\ell_a \geq d \text{ (greater effective depth)} = 7 - 0.75 - 0.25 = 6 \text{ in.}$$

$$\geq 12d_b = 12 \times 0.5 = 6 \text{ in.}$$

$$\geq \tfrac{1}{16}\,\ell_n = 19 \times \tfrac{12}{16} = 14.25 \text{ in.}$$

Therefore, the bars need to extend $57 + 14.25 = 72$ in. $= 6$ ft beyond the face of the support.

Additional checks: We have already checked for deflection and for minimum reinforcement. The maximum spacing of the reinforcement must not exceed (Section 6.4)

$$3 \times 7 = 12 \text{ in.} \quad \text{or} \quad 18 \text{ in.} \quad \text{O.K.}$$

The weight of a 7-in.-thick reinforced concrete slab is (Section 7.1)

$$7 \times 12 = 84 \text{ psf}$$

which allows $100 - 84 = 16$ psf for finishes. O.K.

It is not necessary to check a beam-supported slab for shear. The reinforcement ratio ρ is less than 0.005, and the possibility that the concrete may be overstressed need not therefore be checked (see Table 9.5).

Example 12.1B

Design the corner panel of a continuous floor structure consisting of beams and two-way slabs, spanning 6 m between beam centers in both directions. The dead load is 5 kPa, the live load is 3 kPa, the concrete is Grade 25, and the steel is Grade 400.

Data given: $D = 5$ kPa, $L = 3$ kPa, $\ell = 6$ m, $b = 1$ m, $f'_c = 25$ MPa, $f_y = 400$ MPa, and $\phi = 0.90$.

The ultimate load is

$$w_u = 1.4 \times 5 + 1.7 \times 3 = 12.1 \text{ kPa}$$

Let us assume that the supporting beams are 300 mm wide, so that the clear span is

$$\ell_n = 5\ 700 \text{ mm}$$

Minimum depth: Control of deflection limits the overall depth (see Section 8.7). From Table 8.3, $h \geqslant 90$ mm. If f_y is in MPa, Eq. (8.5)* becomes

$$h \geqslant \frac{\ell_n}{36\ 000}\left(800 + \frac{f_y}{1.5}\right) = \frac{5\ 700 \times 1\ 067}{36\ 000} = 169 \text{ mm}$$

Use a 175-mm slab.

Minimum reinforcement: The minimum amount of reinforcement to resist temperature and shrinkage stresses (Section 6.3) is

$$A_s = 0.001\ 8\ bh = 0.001\ 8 \times 1\ 000 \times 175 = 315 \text{ mm}^2/\text{m}$$

This requires No. 10 bars on no more than 300-mm centers.

Reinforcement required by bending moment: The reinforcement in one direction must lie above the reinforcement in the other direction. Assuming 10-mm bars both ways, the lesser effective depth is

$$d = 175 - 20 \text{ (cover)} - 10 \text{ (outer bars)} - \tfrac{1}{2} \times 10 \text{ (inner bars)} - 140 \text{ mm}$$

Let us assume conservatively (Table 9.1) that

$$z = 0.90d = 126 \text{ mm}$$

Using the direct method, the maximum bending moments in the middle strip, from Table 12.1, are

$$\text{Negative moment at continuous edge} = 0.049w_u\ell_s^2 = 0.049 \times 12.1 \times 6^2$$

$$= 21.34 \text{ kN} \cdot \text{m/m}$$

$$\text{Negative moment at discontinuous edge} = 0.025w_u\ell_s^2 = 10.89 \text{ kN/m}$$

$$\text{Positive moment at midspan} = 0.037w_u\ell_s^2 = 16.12 \text{ kN/m}$$

The maximum bending moments in the column strip are two-thirds of these values.

The absolute maximum bending moment requires, from Eq. (9.8), a steel area of

$$A_s = \frac{M_u}{\phi f_y z} = \frac{21.34 \times 10^{-3}}{0.9 \times 400 \times 0.126} = 470 \times 10^{-6} \text{ m}^2 = 470 \text{ mm}^2$$

Use No. 10 bars on 200-mm centers at continuous edge ($= 500$ mm²/m).

---

*This is Eq. (9.13) of the *ACI Code*; h could be reduced to 140 mm (giving a 150-mm slab) by using the more complicated Eq. (9.11) of the *ACI Code*.

Selection of reinforcement: Construction would become complicated if we used in the middle strip in conjunction with the 200-mm spacing any spacing other than 200 mm or 400 mm for the other bending moments. Furthermore, there are no bars smaller than No. 10. But the minimum area of reinforcement calculated above is No. 10 at 300 mm. Therefore, it is appropriate to use the 200-mm spacing also for the lower bending moments throughout the middle strip. *Use No. 10 bars on 200-mm centers throughout the middle strip.*

In the column strip bending moments are reduced by one-third, and we require a steel area of

$$A_s = \frac{2}{3} \times 470 = 313 \text{ mm}^2/\text{m}$$

Use No. 10 bars on 300-mm centers throughout the column strips ($= 333 \text{ mm}^2/\text{m}$).
The middle strip is $\frac{1}{2} \times 5\ 700 = 2\ 850$ mm wide, and thus requires

$$\frac{2\ 850}{200} = 15 \text{ reinforcing bars for each bending moment}$$

The column strips are each 1 425 mm wide and require

$$\frac{1\ 425}{300} = 5 \text{ reinforcing bars}$$

Arrangement of reinforcement: The bars can either be run straight at the top for the negative bending moments, and straight at the bottom for the positive bending moment, or we can use bars bent up alternatively left and right, each type of bar at a 400-mm spacing in the middle strip and at 600-mm spacing in the column strips. The second arrangement requires expensive bar bending, but it uses less steel, and makes it easier to fix the negative reinforcement. This more complicated arrangement is shown in Fig. 12.4 in American units.

12.4 HOLLOW-BLOCK SLAB CONSTRUCTION AND CONCRETE-JOIST SLAB CONSTRUCTION

We noted in Section 10.1 that parts of the tension zone could be cut away from a solid slab, and the reinforcement concentrated in the ribs (see Fig. 10.4) without in any way reducing the strength of the slab (although we would reduce the stiffness). This concept is employed systematically in concrete-joist slab construction [Fig. 12.5(b)].

Alternatively, we can produce ribbed construction with a flat soffit by using hollow blocks of concrete or clay to form the ribs, and leaving them permanently in position. Hollow-block floors and roofs, for the same dimensions, have better thermal insulation and slightly better sound insulation. They are popular in many European countries.

Both types of construction can be used in one-way slabs (Section 12.1), in two-way slabs (Section 12.2), or with two-way ribs in flat slabs or flat plates (Section 12.5). An example of a flat plate with concrete-joist slab construction is shown in Fig. 12.7.

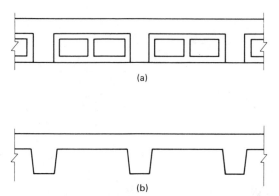

Figure 12.5   (a) Hollow-block slab construction.   (b) Concrete-joist slab construction.

When hollow blocks are used, the walls of the block (if of suitable strength and stiffness) in contact with the ribs may be considered as contributing both to the resistance moment and to the shear resistance of the structure.

12.5 FLAT SLABS AND FLAT PLATES

A *flat slab* is defined as a concrete slab carried directly on enlarged column heads without beams.   The enlarged column heads reduce the shear stress around the columns, and they are quite attractive in appearance for garages, and similar buildings where the column heads give an appearance of strength. In residential buildings, such as blocks of apartments or condominiums, the enlarged column heads would project from the walls or partitions, and in office buildings they would generally be too deep to be accommodated within a false ceiling.   This can be overcome by omitting the column heads.   Slabs supported directly on the columns without column heads are called *flat plates* [Fig. 12.6(a)].

Flat plates can be used with thick slabs lightened by coffers (Fig. 12.7), with the reinforcement concentrated in the ribs between the coffers, over comparatively large spans, so that a wide column spacing permits ample room for the maneuvering of vehicles.   Flat plates are more commonly used with quite thin solid slabs, that is, the minimum permitted by the rules cited in Section 8.7, over relatively small spans for residential buildings and small commercial buildings without air conditioning where false ceilings are not needed.   When false ceilings are required, beam-supported slabs are often cheaper because they have a greater overall depth, and therefore permit larger spans with less concrete in the floor structure.   The formwork is very simple, so that the construction is relatively cheap, and no finishes are required beyond, say, a sprayed acoustic plaster on the ceiling, and a covering on the floor.   Creep deflection is, however, a special problem in flat plates, and this

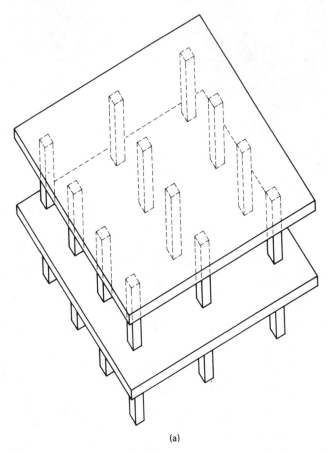

(a)

Figure 12.6   Flat plates (a) consist of reinforced concrete slabs supported directly on reinforced concrete columns.   They are, from a constructional point of view, the simplest form of reinforced concrete structures.   Flat slabs (b) have, in addition, enlarged column heads.

is the reason for keeping the spans small.   In residential buildings this is not a serious handicap.

Flat slabs were patented by Orlando Norcross in the United States in 1902.   The earliest design methods, due mainly to C. A. P. Turner, were purely empirical.   In the years following World War I, H. M. Westergaard produced a theoretical solution.   This overestimates the moments because of stress redistribution between adjacent panels (see Section 8.2), and a modified method, based on an experimental investigation (Ref. 12.1), was first introduced into the American concrete code in 1925.   With some modifications, this is the *direct method* of Section 13.6 of the *ACI Code*.   The method is subject to the restraints set out in Section 13.6.

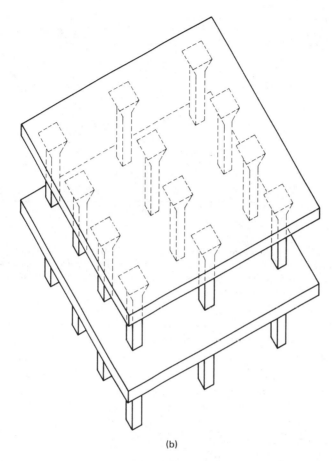

(b)

Figure 12.6   (*Continued*)

The flat plate in Example 12.2 is designed in accordance with the bending moment coefficients of this direct design method.   This is too long to reproduce here; please refer to the *ACI Code* if you wish to follow Example 12.2.

In the alternative *equivalent frame method* the flat slab or flat plate and its supporting columns are considered as a series of elastic frames at right angles to one another.   Each of these frames can be analyzed as if it were a frame consisting of beams and columns.

Only square or nearly square panels are permitted in flat-slab or flat-plate construction.   We noted in Section 12.2 that in a beam-supported square two-way slab, half the load is carried in each direction.   In the design of flat slabs and flat plates by the equivalent frame method the *full* load is carried in each of the two directions at right angles to one another.   Because of the absence of beams, the slab performs a double function: it spans as a slab

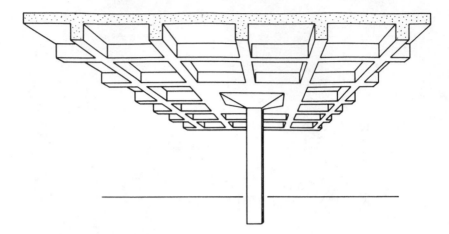

Figure 12.7 Flat plate floor formed with two-way ribs. A solid slab is used in the portion surrounding the column, partly to resist the high shear stresses around the column, and partly to increase the area of concrete in compression in the region of negative bending moment.

between the column strips, and then as a shallow beam between the columns. Thus the bending moment in a square flat plate or flat slab is in each direction approximately the same as that in a beam-supported slab spanning in one direction only, assuming that the beams are strong enough to transmit the reaction of the floor slab to the columns.

The equivalent frame analysis can be made with a suitable computer program (see Section 7.3) or by moment distribution. Tables of moment distribution constants for flat slabs and flat plates are given in the *ACI Code Commentary* (Ref. 4.5, pp. 100–105).

Whatever method is used, the *shear around the column* must be checked, and this is often critical for the thickness of the slab. For a *flat plate*, the shear force (Section 11.2) is divided by the effective depth d and by the perimeter b_0 around the column at a distance $\frac{1}{2}d$ from the column face (Fig. 12.8). Thus the nominal ultimate shear stress

$$v_n = \frac{V_n}{b_0 d} \tag{12.3}$$

This shear is different in character from the flexural shear that was

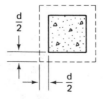

Figure 12.8 The critical section for shear around a column supporting a flat plate is the perimeter b_0 that runs at a distance of $\frac{1}{2}d$ from the column face. Note that the distance is $\frac{1}{2}d$, not d as in Fig. 11.6.

considered in Chapter 11. The column tends to punch through the slab, leaving a hole with a 45° edge. Tests on flat slabs and flat plate structures show that the shear strength of the concrete is higher under these conditions, and Section 11.11 of the *ACI Code* limits the concrete stress to $4\sqrt{f_c'}$, that is, twice the limit for flexural shear.

If v_n calculated from Eq. (12.3) exceeds $4\sqrt{f_c'}$, the plate must be reinforced, or it must be made thicker, or a column with a larger circumference must be used. It is generally simplest to increase the thickness of the slab.

In a *flat slab* the perimeter b_0 is formed at a distance $\frac{1}{2}d$ from the edge of the enlarged column head. Because of the much greater shear perimeter, a flat slab has a higher shear capacity than a flat plate.

If the shear force is high, a *drop panel* can be used. The soffit is dropped 2 to 4 in. to give a greater slab thickness for a rectangular panel around the column, which has a width equal to about one-fourth of the span in both directions.

The shear stress around the column head is reduced by the increased depth. A further check for shear must be made at the edge of the drop panel; however, since b_0 is now the circumference of the drop panel, this is not likely to be critical.

12.6 FLAT PLATE EXAMPLE

Example 12.2A

Design an interior panel of a flat plate structure, spanning 20 ft between column centers in both directions. The columns are 12-in. square. The dead load is 120 psf, the live load is 60 psf, the concrete is Grade 4, and the steel is Grade 60.

Data given: $D = 120$ psf, $L = 60$ psf, $\ell = 20$ ft, $\ell_n = 19$ ft $= 228$ in., $b = 1$ ft, $f_c' = 4000$ psi, $f_y = 60,000$ psi, and $\phi = 0.90$ for bending and 0.85 for shear.

The ultimate load is

$$w_u = 1.4 \times 120 + 1.7 \times 60 = 270 \text{ psf}$$

Depth by deflection limit: Control of deflection may limit the overall depth (see Section 8.7). From Table 8.3, $h \geq 5$ in. From Eq. (8.5),

$$h \geq \frac{\ell_n}{36\,000}\left(800 + \frac{f_y}{200}\right) = \frac{228 \times 1100}{36\,000} = 6.97 \text{ in.}$$

Depth by shear resistance around column: Shear usually determines the depth of a flat plate structure. It must be checked at a square section $\frac{1}{2}d$ from the face of the column. To perform the calculation, we must assume a slab depth h and an effective depth d that err on the safe side; that is h and d must not be greater than those calculated below.

We will assume that $h = 8$ in. and take the effective depth for the purpose of computing the shear around the column conservatively as the smaller effective depth of the inner layer of steel.

$$d = 8 - 0.75 \text{ (cover)} - 0.5 \text{ (bar diameter)} - \tfrac{1}{2} \times 0.5 \text{ (bar radius)} = 6.5 \text{ in.}$$

Therefore, the side of the square shown dashed in Fig. 12.8 is

$$12 + 6.5 = 18.5 \text{ in.} = 1.54 \text{ ft}$$

The ultimate shear force per column is the load acting on the flat panel 20 ft square, minus the load carried by the square around the column.

$$V_{u.} = 270 \,(20^2 - 1.54^2) = 107,360 \text{ lb}$$

and the nominal ultimate shear force is

$$V_n = \frac{V_u}{\phi} = \frac{107,360}{0.85} = 126,300 \text{ lb}$$

The circumference on which this shear force acts is

$$b_0 = 4 \times 18.5 = 74 \text{ in.}$$

and the limiting shear stress around the column of a flat plate is

$$4\sqrt{f_c'} = 253 \text{ psi}$$

Therefore, the effective depth needed for shear, from Eq. (12.3), is

$$d = \frac{V_n}{b_0 \, 4\sqrt{f_c'}} = \frac{126,300}{74 \times 253} = 6.75 \text{ in.}$$

That is more than the depth obtained by the deflection criterion. The overall depth is

$$h = 6.75 + 0.75 + 0.5 + 0.25 = 8.25 \text{ in.}$$

Use an 8½-in slab.

Minimum reinforcement: The minimum amount of reinforcement to resist temperature and shrinkage stresses (Section 6.3) is

$$A_s = 0.0018 \, bh = 0.0018 \times 12 \times 8.5 = 0.184 \text{ in.}^2/\text{ft}$$

which requires No. 4 bars at 12-in. centers.

Reinforcement required by bending moment: The bending moments for a square flat plate designed by the direct method are derived from a basic moment defined in Section 13.6.2 of the *ACI Code:*

$$M_0 = \tfrac{1}{8} \, w_u \ell \ell_n^2$$

$$= \tfrac{1}{8} \times 270 \times 20 \times 19^2 = 243,675 \text{ lb·ft}$$

$$= 2,924,000 \text{ lb·in.}$$

distributed over the entire width of the slab.

Sections 13.6.3 and 13.6.4 of the *ACI Code* explain how this moment is distributed between the column and middle strips, and between positive and negative moments. The highest single bending moment is the negative moment in the column strip, as may be expected, since that determines the reinforcement directly above the column.

The *ACI Code* gives 65% of M_0 to the total negative moment and 35% to the total positive moment.   It gives 75% of M_0 to the column strip and 25% to the middle strip.   Therefore, the highest bending moment is

$$0.65 \times 0.75 \times 2,924,000 = 1,425,000 \text{ lb} \cdot \text{in.}$$

This acts on the width of the column strip, which is ½ × 20 = 10 ft.

Assuming No. 5 bars, the effective depth of the inner layer of reinforcement is

$$d = 8.5 - 0.75 - 0.63 - 0.31 = 6.81 \text{ in.}$$

Taking z conservatively as $0.90d$ (Table 9.1), we obtain

$$z = 6.13 \text{ in.}$$

Therefore, the area of reinforcement required for the 10 ft width of the column strip is

$$A_s = \frac{M_u}{\phi f_y z} = \frac{1,425,000}{0.90 \times 60,000 \times 6.13} = 4.305 \text{ in.}^2$$

That equals

$$\frac{4.305}{10} = 0.431 \text{ in.}^2/\text{ft}$$

for a unit width, which is more than the minimum reinforcement calculated above.

The next highest bending moment is the positive moment in the column strip, which resists the remaining 35% of the moment in the column strip:

$$0.35 \times 0.75 \times 2,924,000 = 768,000 \text{ lb} \cdot \text{in.}$$

This requires a steel area of

$$A_s = \frac{768,000}{0.9 \times 60,000 \times 6.13} = 2.320 \text{ in.}^2$$

which equals

$$\frac{2.320}{10} = 0.232 \text{ in.}^2/\text{ft}$$

This also is higher than the minimum area of reinforcement, namely 0.184 in.^{2}/ft.

The next highest bending moment is the negative moment in the middle strip, which is $0.65 \times 0.25M_0$.   This evidently produces a steel area less than the minimum required, as does the even smaller positive moment in the middle strip.   We thus require the following reinforcement (in in.^{2}/ft).

	Middle Strip	Column Strip
Positive reinforcement	0.184	0.232
Negative reinforcement	0.184	0.431

Use No. 4 bars at 12-in. centers (= 0.20 in.²/ft) in the middle strip for both positive and negative reinforcement.   Ten bars are required in the 10-ft width for each of the positive and the negative reinforcements.   *Use No. 5 bars at 7½-in. centers (=* *0.50 in.²/ft) in the column strip for negative reinforcement and No. 5 bars at 15-in.* *centers (= 0.25 in.²/ft) for positive reinforcement.*   In the 10-ft width, 16 bars are required for the negative reinforcement and 8 bars for the positive reinforcement.

Arrangement of the reinforcement: The reinforcement can be used in straight lengths or bent-up bars.   In Example 12.1 we used bent-up bars; in this example we will use straight bars instead to illustrate the difference.   The resulting layout is shown in Fig. 12.9.   Figure 13.4.8 of the *ACI Code* (Ref. 4.1) sets out the limitations on the bending up and curtailment of bars, which are mainly determined by the requirements for adequate anchorage of the bars.   This is reproduced in Fig. 12.10.

The reinforcement is provided by 7 bar types.   Bars A and B provide the positive reinforcement in the middle strip, bars C and D the positive reinforcement in the column strip, bar E the negative reinforcement in the middle strip, and bars F and G the negative reinforcement in the column strip.   Refer to Figs. 12.9 and 12.10:

5 bars of Type A (No. 4 bars at the bottom of the slab) start $0.15\ell =$ 3 ft from the column face line and project 3 in. beyond the opposite column face line.

5 bars of Type B are similar, but face the opposite way.

4 bars of Type C (No. 5 bars at the bottom of the slab) start $0.125\ell =$ 2 ft 6 in. from the column face line, and project 3 in. beyond the opposite column face line.

4 Type D bars are similar, but face the opposite way.

10 Type E bars (No. 4 bars at the top of the slab) project $0.22\ell_n = 4$ ft 5 in. beyond the column face on either side.

8 Type F bars (No. 5 bars at the top of the slab) project $0.20\ell_n = 4$ ft beyond the column face on either side.

8 Type G bars (No. 5 bars at the top of the slab) project $0.30\ell_n = 6$ ft beyond the column face on either side.

Figure 12.9 shows the reinforcement in one direction only.   The reinforcement in the other direction is precisely the same for a square panel.

Additional checks: The anchorage requirements, set out in Fig. 12.10, were used for the reinforcement layout in Fig. 12.9.   We have also calculated deflection, shear, and minimum reinforcement.

The maximum spacing of the reinforcement must not exceed (Section 6.4)

$$3 \times 8.5 = 25.5 \text{ in.}   \text{ or }   18 \text{ in.}   \text{O.K.}$$

The weight of an $8\frac{1}{2}$-in.-thick reinforced concrete slab is (Section 7.1)

$$8.5 \times 12 = 102 \text{ psf}$$

which allows 18 psf for finishes.

The reinforcement ratio ρ is less than 0.005, and the possibility that the concrete may be overstressed need not be considered (see Table 9.5).

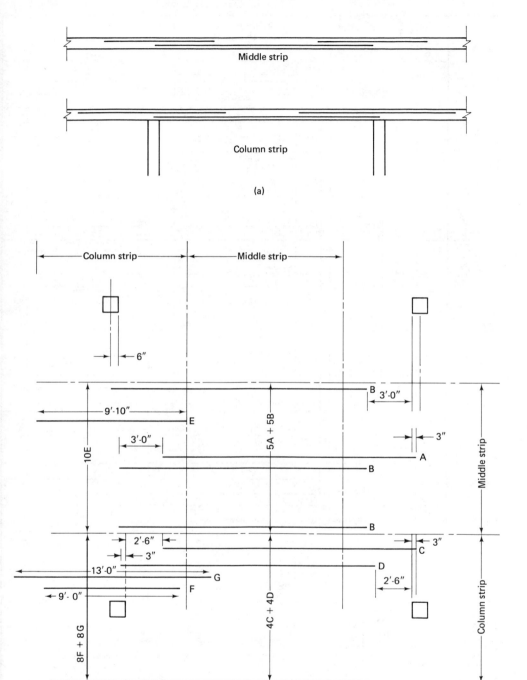

Figure 12.9 (a) Cross section and (b) plan of one interior panel of the flat-slab structures designed in Example 12.2.A. Only reinforcing bars in one direction are shown. The bars in the other direction are identical, only their effective depth differs by one bar diameter.

193

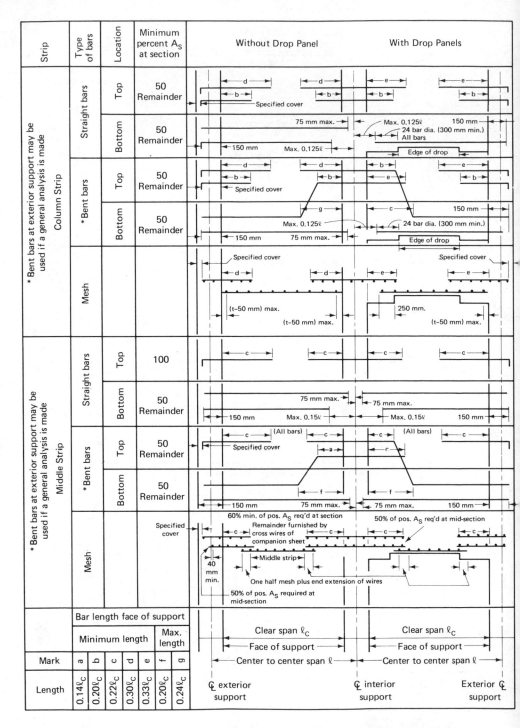

Figure 12.10 Minimum length of reinforcement in flat slabs and flat plates, from Fig. 13.4.8 of the *ACI Code* (Ref. 4.1.).

Example 12.2B

Design an interior panel of a flat plate structure, spanning 6 m between column centers in both directions. The columns are 300-mm square. The dead load is 6 kPa, the live load is 3 kPa, the concrete is Grade 25, and the steel is Grade 400.

Data given: D = 6 kPa, L = 3 kPa, ℓ = 6 m, ℓ_n = 5.7 m, b = 1 m, f_c' = 25 MPa, f_y = 400 MPa, and ϕ = 0.90 for bending and 0.85 for shear.

The ultimate load is

$$w_u = 1.4 \times 6 + 1.7 \times 3 = 13.5 \text{ kPa}$$

Depth by deflection limit: Control of deflection may limit the overall depth (see Section 8.7). From Table 8.3, $h \geq 120$ mm. If f_y is in MPa, Eq. (8.5) becomes

$$h \geq \frac{\ell_n}{36\ 000}\left(800 + \frac{f_y}{1.5}\right) = \frac{5\ 700 \times 1\ 067}{36\ 000} = 169 \text{ mm}$$

Depth by shear resistance around column: Shear usually determines the depth of a flat plate structure. It must be checked at a square section $\frac{1}{2}d$ from the face of the column. To perform the calculation, we must assume a slab depth h and an effective depth d that err on the safe side; that is, h and d must not be greater than those calculated below.

We will assume that h = 200 mm, and take the effective depth for the purpose of computing the shear around the column conservatively as the smaller effective depth of the inner layer of steel.

$$d = 200 - 20 \text{ (cover)} - 10 \text{ (bar diameter)} - \frac{1}{2} \times 10 \text{ (bar radius)} = 165 \text{ mm}$$

Therefore, the side of the square shown dashed in Fig. 12.8 is

$$300 + 165 \text{ mm} = 465 \text{ mm}$$

The ultimate shear force per column is the load acting on the flat plate panel 6 m square, minus the load carried by the square around the column.

$$V_u = 13.5(6.000^2 - 0.465^2) = 483.1 \text{ kN}$$

and the nominal ultimate shear force is

$$V_u = \frac{V_u}{\phi} = \frac{483.1}{0.85} = 568.3 \text{ kN}$$

The circumference on which this shear force acts is

$$b_0 = 4 \times 465 = 1860 \text{ mm}$$

and the limiting shear stress around the column of a flat plate is

$$\tfrac{1}{3}\sqrt{f_c'} = 1.67 \text{ MPa}$$

Therefore, the effective depth needed for shear, from Eq. (12.3), is

$$d = \frac{V_n}{b_0 \, \tfrac{1}{3}\sqrt{f_c'}} = \frac{568.3 \times 10^{-3}}{1.86 \times 1.67} = 0.184 \text{ m}$$

That is more than the depth obtained by the deflection criterion. The overall depth is

$$h = 184 + 5 + 10 + 20 = 219 \text{ mm}$$

Use a 225-mm slab.

Minimum reinforcement: The minimum amount of reinforcement to resist temperature and shrinkage stresses (Section 6.3) is

$$A_s = 0.001\ 8\ bh = 0.001\ 8 \times 1\ 000 \times 225 = 405 \text{ mm}^2/\text{m}$$

which requires No. 10 bars on 200-mm centers, or No. 15 bars on 400-mm centers.

Reinforcement required by bending moment: The bending moments for a square flat plate designed by the direct method are derived from a basic moment defined in Section 13.6.2 of the *ACI Code:*

$$M_0 = \tfrac{1}{8}\ w_u \ell \ell_n^2$$

$$= \tfrac{1}{8} \times 13.5 \times 6.0 \times 5.7^2 = 329.0 \text{ kN} \cdot \text{m}$$

distributed over the entire width of the slab.

Section 13.6.3 and 13.6.4 of the *ACI Code* explain how this moment is distributed between the column and middle strips, and between positive and negative moments. The highest single bending moment is the negative moment in the column strip, as may be expected, since that determines the reinforcement directly above the column.

The *ACI Code* gives 65% of M_0 to the total negative moment and 35% to the total positive moment. It gives 75% of M_0 to the column strip and 25% to the middle strip. Therefore, the highest bending moment is

$$0.65 \times 0.75 \times 329.0 = 160.4 \text{ kN} \cdot \text{m}$$

This acts on the width of the column strip, which is $\frac{1}{2} \times 6 = 3$ m.

Assuming No. 15 bars, the effective depth of the inner layer of reinforcement is

$$d = 225 - 20 - 5 - 8 = 182 \text{ mm}$$

Taking z conservatively as $0.90d$ (Table 9.2), we obtain

$$z = 164 \text{ mm}$$

Therefore, the area of reinforcement required for the 3-m width of the column strip is

$$A_s = \frac{M_u}{\phi f_y z} = \frac{160.4 \times 10^{-3}}{0.90 \times 400 \times 0.164} = 2.717 \times 10^{-3} \text{ m}^2$$

That equals

$$\frac{2\ 717}{3} = 906 \text{ mm}^2/\text{m}$$

for a unit width, which is more than the minimum reinforcement calculated above.

The next highest bending moment is the positive moment in the column strip, which resists the remaining 35% of the moment in the column strip:

$$0.35 \times 0.75 \times 329.0 = 86.4 \text{ kN} \cdot \text{m}$$

This requires a steel area of

$$A_s = \frac{86.4 \times 10^{-3}}{0.9 \times 400 \times 0.164} = 1.463 \times 10^{-3} \text{ m}^2$$

which equals

$$\frac{1\ 463}{3} = 488 \text{ mm}^2/\text{m}$$

This also is higher than the minimum area of reinforcement, namely 405 mm²/m.

The next highest bending moment is the negative moment in the middle strip, which is $0.65 \times 0.25M_0 = 53.5$ kN·m. This evidently produces a steel area less than the minimum required, as does the even smaller positive moment in the middle strip. We thus require the following reinforcement (in mm²/m):

	Middle Strip	Column Strip
Positive reinforcement	405	488
Negative reinforcement	405	906

Use No. 10 bars on about 200-mm centers (= 500 mm²/m) *in the middle strip for both positive and negative reinforcement.* Fifteen bars are required for the 3-m width. *Use No. 15 bars on about 200-mm centers* (= 1 000 mm²/m) *in the column strip for negative reinforcement and No. 15 bars on about 400-mm centers* (= 500 mm²/ m) *for positive reinforcement.* In the 3-m width, 15 bars are required for negative reinforcement, and 8 bars for positive reinforcement.

Figure 12.9 shows the arrangement of the reinforcement in one direction only in American units.

EXERCISES

1. How does the decision to support a slab directly on the columns without beams affect (a) the construction and (b) the design of the structure?

2. Distinguish between one-way slabs and two-way slabs. Why is a two-way slab system possible only for near-square slabs, that is, slabs whose longer side is less than twice the length of the shorter side?

3. How do you determine the overall depth (a) of a one-way slab and (b) of a two-way slab?

4. How do you determine the amount of reinforcement required in both directions (a) in a one-way slab and (b) in a two-way slab?

5. Distinguish between flat slabs and flat plates.

6. Explain why column heads and drop panels reduce the overall thickness of beam-less slabs.

7. Describe hollow-block slab construction and concrete-joist slab construction. Why is the portion around the column head frequently left solid in both types of construction?

8. Explain the meaning of column strip and middle strip. How wide are the column and middle strips (a) in two-way slabs, (b) in flat slabs, and (c) in flat plates?

9. Why does the *ACI Code* require a greater minimum thickness for flat slabs and flat plates than for two-way slabs?

10. Why is the thickness of two-way slabs normally determined by the requirement to control deflection, whereas the thickness of flat plates is normally determined by the limiting shear stress? Why does the magnitude of the bending moment have relatively little influence on thickness in either case?

11. A two-way slab spans 18 ft between centers of beams, 12 in. wide, in both directions. It carries a dead load of 95 psf and a live load of 60 psf. Determine the minimum thickness of the slab using Grade 3.5 concrete and Grade 60 reinforcement.

12. A flat plate, supported on 12-in.-square columns, spans 18 ft in both directions. It carries a dead load of 110 psf and a live load of 60 psf. Determine its minimum thickness using Grade 3.5 concrete and Grade 60 reinforcement.

REFERENCE

12.1. H. M. Westergaard and W. A. Slater, "Moments and Stresses in Slabs," *Proc. American Concrete Institute*, Vol. 17, 1921, pp. 415–538.

Chapter 13

Design of Columns

We design some simple concentrically loaded interior columns, both square and circular. The design of eccentrically loaded columns, which includes all exterior columns, is more complicated, and discussed only in general terms. Brief mention is made of corner columns, slender columns, spiral reinforcement, seismic design, and reinforced concrete walls.

13.1 COLUMN LAYOUT AND GENERAL DESIGN DETAILS

The layout of the columns is largely determined by the height of the building. In a low building the bending moments due to the wind loads are small, and thus the columns are predominantly in compression. In a tall building the bending moments normally have a greater influence on the design of the columns. The strength of members in pure compression is proportional to $b \cdot h$ (where b is the width and h is the overall depth), and thus the strength of a column measuring $b \cdot 2h$ is the same as the strength of two columns measuring bh. When bending predominates (see Section 13.8) the strength is almost proportional to bd^2 (where d is the effective depth), and thus the strength of a column measuring $b \cdot 2d$ is double the strength of two columns measuring bd. The argument for a few large columns is thus stronger in the case of tall buildings.

In low buildings, particularly if they are of flat-plate construction, deflection of the floors tends to be a problem (see Sections 8.7 and 12.5), and thus short spans and relatively close column spacing is advisable. In tall buildings thicker floor construction, which deflects less, is appropriate for lateral stability, and thus columns tend to be fewer and of larger cross section. A few large columns occupy less space than a larger number of smaller columns with the same load-bearing capacity. Even so, columns in the interior take up valuable floor space, and this is a particular problem with tall buildings.

There is an increasing tendency for tall buildings to use columns on the outside only where they can be made large and project beyond the facade, leaving an interior space unobstructed by columns.

This new approach has largely invalidated one of the earlier objections to reinforced concrete construction (see Section 1.7), namely that the size of the columns makes reinforced concrete unsuitable for tall buildings. It is also responsible for the declining popularity of the use of spiral reinforcement (see Section 13.6), or steel sections encased in concrete, to reduce the size of the columns.

13.2 REINFORCEMENT DETAILS

We mentioned cover in Section 6.1. For interior columns, the *minimum cover* is 1½ in. (40 mm). For exterior columns, it is 2 in. (50 mm) since longitudinal bars will always be larger than No. 4 (No. 15, metric).

The *minimum number* of bars is 4 when rectangular ties are used, and 6 if circular ties are used.

The *minimum total reinforcement ratio*, ρ_t, is $0.01A_g$ (where A_g is the area of the gross cross section of the concrete column), and the *maximum reinforcement ratio* is $0.08A_g$.

The relatively slender reinforcing bars cannot be made longer than one story height. Where the bars are spliced, the reinforcement is doubled by the overlap. The length of splices is discussed in Section 6.8.

In order to run the bars for the upper column on the same alignment as the bars of the lower column to which they are spliced, one set of bars is offset to form the lap splice; usually, the bars are offset inward at the top and the new set of bars is slid over them from the top. It has been customary to make the splice just above the column–slab junction. However, the problems created by crowding at the junction may be more serious, and in some cases it is better to move the splice higher up, say to one-fourth of the column height. Alternatively, welded or mechanical splices may be used. When columns are subjected to large bending moments, the splices should be at the points of inflection, where there is zero bending moment.

In many columns some of the bars are always in tension because of a bending moment, or some of the bars may be in tension under certain conditions of wind or earthquake loading.   A longer splice is then required (Section 6.8).

The *minimum clear spacing* of longitudinal reinforcement (Section 7.6 of the *ACI Code*) must not be less than 1½ in. (40 mm) or 1½ times the diameter of the largest bar.   This also applies to the clear spacing between a contact lap splice and adjacent splices or bars.

Longitudinal bars are liable to buckle sideways under action of quite small loads unless properly restrained (Fig. 13.1).   Although the concrete provides some restraint, ties are needed at frequent intervals to prevent the bars from bursting the cover.

The *minimum size* of ties (Section 7.10 of the *ACI Code* is No. 3 (No. 10, metric), and the *maximum spacing* is the least of:

48 times the diameter d_b of the tie bar

16 times the diameter d_b of the smallest longitudinal bar

The smallest dimension of the section, h or b

Ties must be arranged so that every corner and alternate longitudinal bar has lateral support provided by the corner of a tie, normally bent to a right angle, but no more than 135°.   If the bars are further than 6 in. apart,

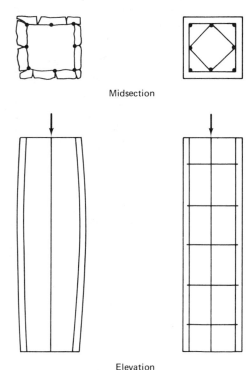

Midsection

Elevation

Figure 13.1 Ties are required at frequent intervals to prevent buckling of the longitudinal bars and consequent bursting of the cover.

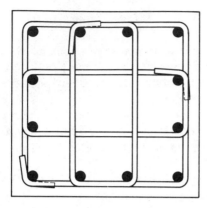

Figure 13.2 Suitable arrangement of column ties for 12 longitudinal bars in a square column, if the bars are more than 6 in. (150 mm) apart.

they must be individually tied. Thus the layout of ties can be quite complex (Fig. 13.2).

When longitudinal bars are arranged in a circle, a complete circular tie may be used, and the bars need not be individually restrained.

Alternatively, *spiral reinforcement* may be used (Fig. 13.3). This greatly adds to the resilience of the column, and it is consequently given a higher strength-reduction factor (see Section 13.3). However, it also adds to the cost, and it is best suited to a circular section.

The *minimum size of spiral reinforcement* (Section 7.10 of the *ACI Code*) is ⅜ in. The maximum center-to-center spacing is 3 in., and the minimum is 1 in.

The rules for seismic design are discussed in Section 13.10.

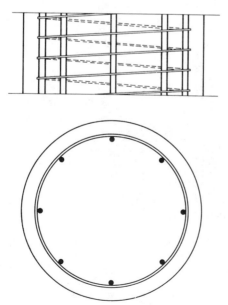

Figure 13.3 Spiral reinforcement surrounding eight longitudinal bars in a circular column.

13.3 STRENGTH OF SHORT, CONCENTRICALLY LOADED COLUMNS

Let us consider a column of gross cross-sectional area A_g (say width b and overall depth h, so that $A_g = bh$), reinforced with a total area of steel (on all faces of the column) A_{st}. The net cross-sectional area of the concrete is $A_g - A_{st}$.

We load the column, and at first both the steel and the concrete behave elastically (Fig. 13.4). At a strain ε_1 we reach the elastic limit of the concrete, and the steel begins to take a greater proportion of the column load. At the strain ε_y we reach the yield or proof strain of the steel, and the steel is thereafter assumed to deform at a constant stress f_y (the yield or proof stress of the steel). The concrete now has to take the whole of the increase in the column load. At the strain ε_3 the concrete reaches its maximum stress $0.85f_c'$, and at this strain the column load cannot be further increased. However, if it is reduced, the column can deform further until the concrete crushes at the strain ε_c', when the column collapses.

The maximum concentric column load P_0 is obtained by multiplying this load by the strength reduction factor ϕ. This is 0.70 for tied columns, and 0.75 for spirally reinforced columns.

$$P_0 = \phi[0.85f_c'(A_g - A_{st}) + f_y A_{st}] \qquad (13.1)$$

This column load P_0 can be attained only if the column is perfectly straight and perfectly concentrically loaded. This is very difficult to ensure,

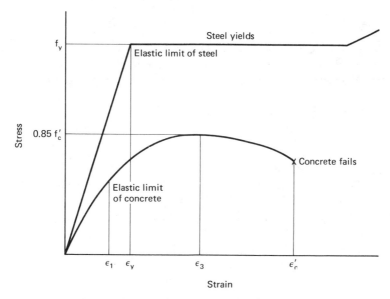

Figure 13.4 Relation between stress and strain for steel and concrete in the hypothetical case of a perfectly concentrically loaded column.

and it is therefore customary to assume that the column has small imperfec-
tions. In the 1963 and 1971 ACI codes this was expressed as a small initial
eccentricity of loading. In the 1977 and 1983 ACI codes, Section 10.3.5.1,
the column load P_0 is instead reduced by 20% for tied columns and by 15%
for spirally reinforced columns. This corresponds approximately to eccentricities of loading of $b/20$ and $b/10$, respectively, where b is the width of a
square column, or the shorter dimension of a rectangular column.

We thus obtain the ultimate load of a short concentrically loaded column:

$$P_u = 0.80\phi[0.85f_c'(A_g - A_{st}) + f_y A_{st}] \tag{13.2}$$

where $\phi = 0.70$, for a tied column, and

$$P_u = 0.85\phi[0.85f_c'(A_g - A_{st}) + f_y A_{st}] \tag{13.3}$$

where $\phi = 0.75$ for a spirally reinforced column. For design, Eq. (13.2) is
more conveniently written in two forms:

$$P_u = 0.56[0.85f_c' (1 - \rho_t) + f_y \rho_t]A_g \tag{13.4}$$

where $\rho_t = A_{st}/A_g$ is the column reinforcement ratio and

$$P_u = 0.56[0.85f_c' A_g + (f_y - 0.85f_c') A_{st}] \tag{13.5}$$

Similar equations can be written in place of Eq. (13.3).

13.4 DESIGN OF RECTANGULAR COLUMNS

Example 13.1A

Design a short concentrically loaded column to carry a dead load of 200,000 lb and
a live load of 80,000 lb, using Grade 4 concrete and Grade 60 steel.

Data given: $D = 200,000$ lb, $L = 80,000$ lb, $f_c' = 4000$ psi, and $f_y = 60,000$
psi.

The ultimate column load is

$$U = 200,000 \times 1.4 + 80,000 \times 1.7 = 416,000 \text{ lb}$$

For a concentrically loaded column, a square cross section is the most economical.
The column is too small to make spiral reinforcement worthwhile. The reinforcement
ratio must lie between 1 and 8%, and for this lightly loaded column we will aim at ρ_t
$= 0.02$ approximately. From Eq. (13.4),

$$416,000 = 0.56[0.85 \times 4000(1 - 0.02) + 60,000 \times 0.02]A_g$$

This gives

$$742,900 = (3332 + 1200)A_g$$

$$A_g = h^2 = 163.9 \text{ in.}^2$$

$$h = 12.80 \text{ in.}$$

We can either use a larger column with less than 2% of steel or a smaller column with more than 2% of steel. *Use a 12 in. by 12. in. column cross section.*
From Eq. (13.5),

$$416,000 = 0.56[0.85 \times 4000 \times 12^2 + (60,000 - 0.85 \times 4000)A_{st}]$$

which gives

$$742,900 = 489,600 + 56,600 \, A_{st}$$

$$A_{st} = 4.475 \text{ in.}^2$$

The column must have at least four bars. There is no anchorage problem in a concentrically loaded column, and therefore no reason to use small bars. On the other hand, the concreting of a vertical column 12 in. square and at least 10 ft long makes it desirable to use as few bars as possible. Therefore, 4 bars are appropriate. *Use 4 No. 10 bars* ($= 5.08$ in.^{2}).
We need No. 3 ties whose spacing is the least of:

$$48 \times 0.375 \text{ (tie diameter)} = 18 \text{ in.}$$

$$16 \times 1.270 \text{ (bar diameter)} = 20.3 \text{ in.}$$

$$\text{column width} = 12 \text{ in.}$$

Use No. 3 ties on 12-in. centers.
The layout of the reinforcement is shown in Fig. 13.5.

Example 13.1B

Design a short concentrically loaded column to carry a dead load of 800 kN and a live load of 400 kN, using Grade 30 concrete and Grade 400 steel.
Data given: $D = 800$ kN, $L = 400$ kN, $f'_c = 30$ MPa, and $f_y = 400$ MPa.
The ultimate column load is

$$U = 800 \times 1.4 + 400 \times 1.7 = 1\,800 \text{ kN} = 1.8 \text{ MN}$$

For a concentrically loaded column, a square cross section is the most economical. The column is too small to make spiral reinforcement worthwhile. The reinforcement ratio must lie between 1 and 8%, and for this lightly loaded column we will aim at $\rho_t = 0.02$ approximately. From Eq. (13.4),

$$1.8 = 0.56[0.85 \times 30(1 - 0.02) + 400 \times 0.02]A_g$$

This gives

$$3.214 = (25.0 + 8.0)A_g$$

$$A_g = h^2 = 0.097\,40$$

$$h = 0.312 \text{ m} = 312 \text{ mm}$$

We can either use a larger column with less than 2% of steel or a smaller column with more than 2% of steel. *Use a 300-mm by 300-mm column cross section.*

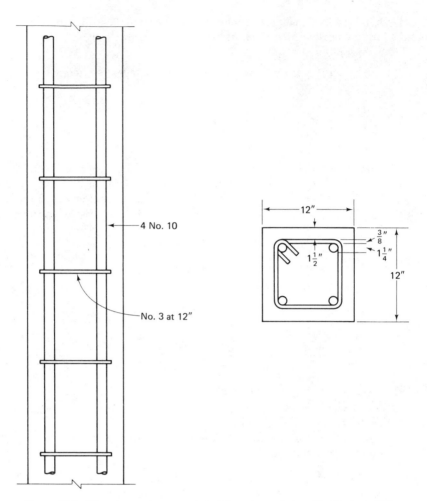

Figure 13.5   Elevation and cross section of the column designed in Example 13.1.A.

From Eq. (13.5),

$$1.8 = 0.56[0.85 \times 30 \times 0.3^2 + (400 - 0.85 \times 30)A_{st}]$$

which gives

$$3.214 = 2.295 + 374.5A_{st}$$

$$A_{st} = 2.454 \times 10^{-3} \text{ m}^2 = 2\ 454 \text{ mm}^2$$

The column must have at least four bars.   There is no anchorage problem in a concentrically loaded column, and therefore no reason to use small bars.   On the other hand, the concreting of a vertical column 300 mm square and at least 3 m long makes it desirable to use as few bars as possible.   Therefore, 4 bars are appropriate. *Use 4 No. 30 bars (= 2 800 mm²).*

We need No. 10 ties whose spacing is the least of:

$$48 \times 10 \text{ (tie diameter)} = 480 \text{ mm}$$

$$16 \times 30 \text{ (bar diameter)} = 480 \text{ mm}$$

$$\text{column width} = 300 \text{ mm}$$

Use No. 10 ties on 300-mm centers.

The layout of the reinforcement is shown in Fig. 13.5 in U.S. units.

Example 13.2

Determine the load-bearing capacity of the column shown in Fig. 13.5, if the concrete is Grade 4 and the reinforcement Grade 60.

Data given: $b = h = 12$ in., $A_s = 4$ No. 10, $f_c' = 4000$ psi, $f_y = 60,000$ psi.

$$A_{st} = 5.08 \text{ in.}^2 \quad \text{and} \quad A_g = 12^2 = 144 \text{ in.}^2$$

From Eq. (13.5),

$$P_u = 0.56[0.85 \times 4000 \times 144 + (60,000 - 0.85 \times 4000) \times 5.08]$$

$$= 435,200 \text{ lb}$$

Example 13.3A

Design a short concentrically loaded column for the ground floor of a four-story building, to carry a dead load of 200,000 lb and a live load of 80,000 lb per floor, using Grade 4 concrete and Grade 60 steel.

Data given: $D = 800,000$ lb, $L = 320,000$ lb, $f_c' = 4000$ psi, and $f_y = 60,000$ psi.

The ultimate column load is

$$U = 800,000 \times 1.4 + 320,000 \times 1.7 = 1,664,000 \text{ lb}$$

We will use a square column with a reinforcement ratio $\rho_t = 0.025$ approximately. From Eq. (13.4),

$$1,664,000 = 0.56[0.85 \times 4000 (1 - 0.025) + 60,000 \times 0.025]A_g$$

$$A_g = h^2 = \frac{2,971,400}{4815} = 617.11$$

$$h = 24.84 \text{ in.}$$

Use a 24 in. by 24 in. cross section.

From Eq. (13.5),

$$A_{st} = \frac{2,971,000 - 0.85 \times 4000 \times 24^2}{60,000 - 0.85 \times 4000} = 17.90 \text{ in.}^2$$

Use 8 No. 14 bars ($= 18.00$ in.^{2}).

Use No. 4 ties, whose spacing is the least of:

$$48 \times 0.5 = 24 \text{ in.}$$

$$16 \times 1.693 = 27.1 \text{ in.}$$

$$\text{column width} = 24 \text{ in.}$$

Use No. 4 ties on 24 in. centers.

Example 13.3B

Design a short concentrically loaded column for the ground floor of a four-story building, to carry a dead load of 800 kN and a live load of 400 kN per floor, using Grade 30 concrete and Grade 400 steel.

Data given: $D = 3\,200$ kN, $L = 1\,600$ kN, $f'_c = 30$ MPa, and $f_y = 400$ MPa. The ultimate column load is

$$U = 3\,200 \times 1.\,4 + 1\,600 \times 1.7 = 7\,200 \text{ kN} = 7.2 \text{ MN}$$

We will use a square column with a reinforcement ratio $\rho_t = 0.025$ approximately. From Eq. (13.4),

$$7.2 = 0.56[0.85 \times 30\,(1 - 0.025) + 400 \times 0.025]A_g$$

$$A_g = h^2 \frac{12.857}{34.86} = 0.368\,8$$

$$h = 0.607 \text{ m}$$

Use a 600-mm by 600-mm cross section.

From Eq. (13.5),

$$A_{st} = \frac{12.857 - 0.85 \times 30 \times 0.6^2}{400 - 0.85 \times 30}$$

$$= 9.818 \times 10^{-3} \text{ m}^2 = 9\,818 \text{ mm}^2$$

Use 4 No. 55 bars ($= 10\,000$ mm²).

Use No. 15 ties, whose spacing is the least of:

$$48 \times 15 = 720 \text{ mm}$$

$$16 \times 55 = 880 \text{ mm}$$

$$\text{column width} = 600 \text{ mm}$$

Use No. 15 ties on 600-mm centers.

13.5 DESIGN OF CIRCULAR COLUMNS

Columns are normally made square or rectangular because the formwork is most easily made that way. When columns are placed in open spaces, a circular cross section often looks more attractive. Circular columns can be formed in nonreusable circular tubes of cardboard or plastic that are peeled off.

Unless they are very heavily loaded, circular columns are designed like rectangular columns.

Example 13.4A

Redesign the column of Example 13.3.A. as a circular column.

Data given: P_u = 1,664,000 lb, f'_c = 4000 psi, and f_y = 60,000 psi.

We will use a reinforcement ratio ρ_t = 0.025 approximately. From Eq. (13.4),

$$1,664,000 = 0.56[0.85 \times 4000\,(1 - 0.025) + 60,000 \times 0.025]A_g$$

$$A_g = \tfrac{1}{4}\pi h^2 = 617.11 \text{ in.}^2$$

$$h = 28.03 \text{ in.}$$

Use a 28-in.-diameter column.

From Eq. (13.5),

$$A_{st} = \frac{2,971,000 - 0.85 \times 4000 \times \tfrac{1}{4}\pi \times 28^2}{60,000 - 0.85 \times 4000} = 15.51 \text{ in.}^2$$

Use 10 No. 11 bars ($= 15.60$ in.^{2}).

Next, check the bar spacing.   The diameter of the bar centers is

$$28 - 2 \times 1.5 \text{ (cover)} - 2 \times 0.5 \text{ (tie)} - 1.41 \text{ (bar diameter)} = 22.59 \text{ in.}$$

On this circle, perimeter $\pi \times 22.59 = 70.96$ in., are 10 bars of 1.41 in. diameter, and 10 spaces between the bars, so that the clear space between bars is

$$\frac{70.96 - 10 \times 1.41}{10} = 5.69 \text{ in.}$$

The minimum admissible spacing is ½ in. or $1\tfrac{1}{2}d_b$ = 2.12 in.   O.K.

We will use No. 4 ties whose minimum spacing is

$$48 \times 0.5 = 24 \text{ in.}$$

$$16 \times 1.41 = 22.6 \text{ in.}$$

$$\text{column diameter} = 28 \text{ in.}$$

Use No. 4 ties at 21-in. centers.

The cross section of the column is shown in Fig. 13.6.

Example 13.4B

Redesign the column of Example 13.3.B. as a circular column.

Data given: P_u = 7.2 MN, f'_c = 30 MPa, and f_y = 400 MPa.

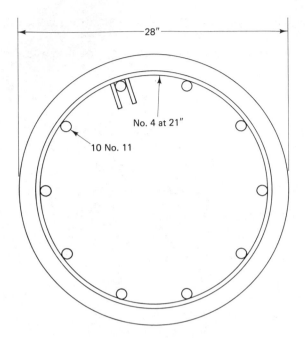

Figure 13.6   Cross section of the column of Example 13.4.A.

We will use a reinforcement ratio $\rho_t = 0.025$ approximately.   From Eq. (13.4),

$$7.2 = 0.56[0.85 \times 30 \ (1 - 0.025) + 400 \times 0.025]A_g$$

$$A_g = \tfrac{1}{4}\pi h^2 = 0.368 \ 8$$

$$h = 0.685 \text{ m}$$

Use a 700-mm diameter column.
   From Eq. (13.5),

$$A_{st} = \frac{12.857 - 0.85 \times 30 \times \tfrac{1}{4}\pi \times 0.7^2}{400 - 0.85 \times 30} = 8.123 \times 10^{-3} \text{ m}^2 = 8 \ 123 \text{ mm}^2$$

Use 12 No. 30 bars (= 8 400 mm²).
   Next, check the bar spacing.   The diameter of the bar centers is

$$700 - 2 \times 40 \text{ (cover)} - 2 \times 15 \text{ (tie)} - 30 \text{ (bar diameter)} = 560 \text{ mm}$$

On this circle, perimeter $\pi \times 560 = 1 \ 760$ mm, are 12 bars of 30 mm diameter, and 12 spaces between the bars, so that the clear space between bars is

$$\frac{1 \ 760 - 12 \times 30}{12} = 116 \text{ mm}$$

The minimum admissible spacing is 40 mm or $1\tfrac{1}{2}d_b = 45$ mm.   O.K.

We will use No. 15 ties whose minimum spacing is

$$48 \times 15 = 720 \text{ mm}$$

$$16 \times 30 = 480 \text{ mm}$$

$$\text{column diameter} = 700 \text{ mm}$$

Use No. 15 ties at 480-mm centers.

13.6 COLUMNS WITH SPIRAL REINFORCEMENT

Spiral reinforcement is useful only for heavily loaded columns of large diameter. It is particularly useful in seismic regions (Section 13.11). The spiral reinforcement, when the column is near failure, applies a lateral pressure to the concrete, and delays its disintegration. The cover often spalls when this happens, and the column strength is therefore based on the cross-sectional area of the column core retained inside the spiral. The spiral reinforcement adds to the strength of the column but increases column strength only for the larger columns, because the spalled concrete cover must be subtracted from the cross-sectional area of the concrete. The formulas are explained in more advanced textbooks on reinforced concrete (e.g., Ref. 7.5, p. 519).

Spirally reinforced columns may be circular or square, but only the circular core of a square column is effective.

13.7 ECCENTRICITY OF COLUMN LOADS, AND BENDING MOMENTS

An ultimate bending moment M_u may be equated to $P_u e$, so that any bending moment resulting from wind loads and other lateral loads can be expressed as an eccentricity of the column load. A deliberate eccentricity occurs commonly in precast construction (Fig. 13.7), but also occasionally in cast-in-place construction for architectural convenience or for the sake of appearance.

Smaller bending moments in columns occur simply as a result of building the columns monolithically with the floor slabs or beams (Fig. 13.8), or because of a change in the disposition of the live load (Fig. 13.9), or because of the slight dimensional inaccuracies which are permitted within the limits of the construction tolerances.

The bending is particularly marked in exterior columns [Fig. 13.10(b)]. Corner columns, in particular, are subject to bending about both axes, which poses a special problem (see Section 13.8). Thus vertical loads can produce bending on columns. In addition, bending is produced by horizontal forces due to wind and due to earthquakes (see Sections 7.1 and 8.6).

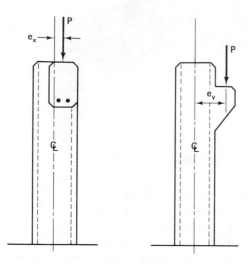

Figure 13.7 Eccentric arrangement of beams in precast construction.

Structures must be designed for the horizontal forces resulting from earthquakes in all earthquake zones (Section 13.11). Wind loads are particularly important in tall buildings and in low buildings not protected by surrounding buildings. If the ultimate bending moment resulting from any of these horizontal forces is M_u, and the ultimate column load is P_u, then the eccentricity of loading resulting from the bending moment is

$$e = \frac{M_u}{P_u}$$

This eccentricity is added to the eccentricity, if any, resulting from the vertical loads, as shown in Figs. 13.7 to 13.10.

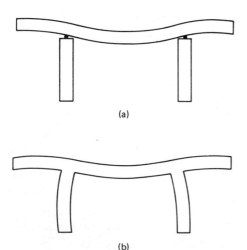

Figure 13.8 Continuous beam. (a) Simply supported on the columns. (b) Monolithic with the columns.

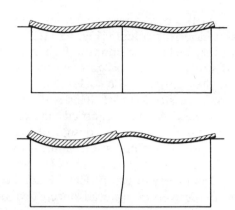

Figure 13.9 Any asymmetry in the load (e.g., removal of the live load on one span) causes bending in a column monolithic with the floor structure.

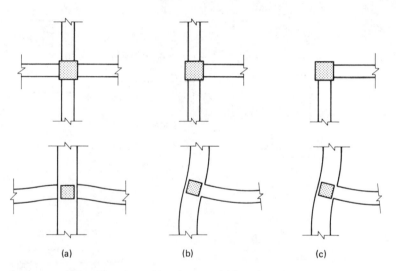

(a)   (b)   (c)

Figure 13.10 Bending of exterior columns. (a) Interior column, no bending under symmetrical loading. (b) Exterior column, bending in one plane. (c) Corner column, bending in both planes.

13.8 THE DESIGN OF ECCENTRICALLY LOADED COLUMNS

An eccentrically loaded column is thus subject to an axial load P_u and a bending moment

$$M_u = P_u e \tag{13.6}$$

When the eccentricity is small, the entire column is in compression, but the compression is higher at one face, and the compressive concrete stress $0.85f_c'$ must not be exceeded.

As the eccentricity increases, tension develops, and the neutral axis passes inside the column [Fig. 13.11(a)]. The steel on one face is in compression; the steel on the opposite face is in tension, but its stress is less than the yield stress. As the eccentricity increases, the steel on one face just starts to yield in tension as the concrete on the opposite face reaches the maximum compressive stress $0.85f_c'$. This is the *balanced failure condition* discussed in Section 9.3. As the eccentricity increases still further, we obtain a primary tension failure, as in the design of beams [Fig. 13.11(b)]. The transition from a primary compression failure to a primary tension failure is shown in Fig. 13.12.

The theory of eccentrically loaded columns leads to complex formulas, that are derived in more advanced textbooks on reinforced concrete design

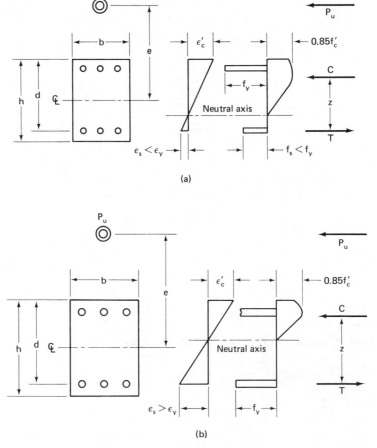

Figure 13.11 Transition from a primary compression failure (a) to a primary tension failure (b) in an eccentrically loaded column.

Figure 13.12 Gradual transition from primary compression failure to primary tension failure as the eccentricity of the column load increases.

(e.g., Ref. 7.5, Chap. 18). Unless a computer is used, they are best solved with the aid of tables or charts, using the *Design Handbook* of the American Concrete Institute for conventional American units (Ref. 13.1). The Canadian Portland Cement Association (Ref. 13.2) has published metric design charts for eccentrically loaded columns conforming to the Canadian Concrete Code (Ref. 4.5), which can be used in conjunction with the metric ACI code.

A separate chart is needed for each concrete and steel stress, and for each reinforcement layout. Reference should be made to more advanced textbooks on reinforced concrete design for examples of the use of charts (Ref. 7.5).

13.9 BIAXIAL BENDING

Many columns are subject to bending about an axis which is not parallel to either face of the column. All corner columns are in that category. There is no simple solution to that problem, but it can be solved approximately by resolving the eccentricity of the column load into two components, e_x and e_y. The column is then designed with the aid of charts, or some other method, as if it has only an eccentricity e_x in the x-direction. This gives the axial load capacity P_x. It is next designed as if it had only an eccentricity e_y in the y-direction. This gives the axial load capacity P_y. The axial load capacity P_0 without any eccentricity is then determined.

The axial load capacity of the column loaded eccentrically in both the

x- and the y-directions is P_u, where

$$\frac{A_g}{P_u} = \frac{A_g}{P_x} + \frac{A_g}{P_y} - \frac{A_g}{P_0}$$

and A_g is the gross area of the column cross section.

13.10 SLENDER COLUMNS

Fortunately, about 90% of all braced columns and about 40% of all unbraced columns can be designed as "short" columns, and therefore slenderness is not frequently a factor in the design of the columns for small buildings. In this respect the design of reinforced concrete columns differs from the design of steel columns.

The design of reinforced concrete columns is, however, more complicated than the design of steel columns. The various methods that may be used for determining the strength of slender columns are discussed in the *ACI Code Commentary* (Ref. 4.5, pp. 46–59) and also in advanced textbooks on reinforced concrete design (e.g., Ref. 7.5, Chap. 19).

The simplest method of allowing for the reduction in column strength due to the buckling sideways of slender columns is the *modified R method*. The short column load is multiplied by a column reduction factor R, which is reduced linearly as the slenderness ratio increases (Ref. 7.5, pp. 562–576).

13.11 SEISMIC DESIGN OF COLUMNS

The most common cause of the collapse of a building in an earthquake, or of major damage to it, is insufficient ductility in the bearing walls and columns. An earthquake is a rapid movement of the ground, which frequently lasts for less than a minute. The foundation of the building moves with the ground, but because of its inertia the building is left behind. The vertical component of this movement is easily resisted by most buildings, because they are designed to resist large vertical forces. The horizontal component causes the damage, unless the building has been designed to resist the horizontal forces.

Traditional structures of natural stone, brick, blocks, or consolidated soil have little or no ductility and are liable to be damaged even by small earthquakes. Steel structures have excellent ductility, and ductility can be introduced into reinforced concrete structures by suitable reinforcement. The ductility of reinforced concrete is entirely a function of its reinforcement, which must be provided in every direction where tension may occur.

The *ACI Code* in Appendix A on Seismic Design recommends that

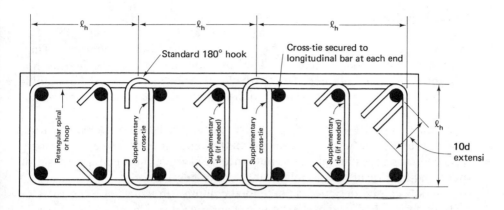

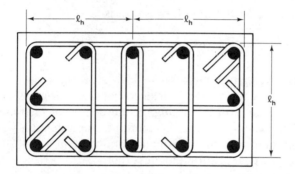

Figure 13.13 Supplementary ties for columns in seismic zones. [From *ACI Code Commentary* (Ref. 4.3).]

additional hoops or supplementary ties (Fig. 13.13) be used to restrain the longitudinal column bars. The ties provide shear reinforcement to resist any shear forces on the columns produced by earthquake loads. They also compensate for spalling of the cover, should it occur. The use of spiral reinforcement (Section 13.6) is even more effective.

The extra ties make possible the formation of plastic "hinges" without disintegration of the concrete if the earthquake forces are greater than anticipated. It is, however, undesirable for plastic hinges to form in the columns because they endanger the stability of the entire frame. At all column-to-beam connections, the moment strengths of the columns should therefore exceed those of the beams, so that plastic hinges form first in the beams.

Reference has already been made to the use of tension and compression reinforcement in flexural members, and to shear walls (Sections 9.3, 10.7, and 11.6).

13.12 WALLS

In traditional construction walls were almost invariably load-bearing. They
were thus part of the structure. The resulting thick walls also had an important environmental role. They provided thermal inertia, which kept buildings cool on a hot day and during a cold night delayed the cooling of the
interior.

Since the early years of this century, curtain walls without a load-bearing
function, and sometimes also transparent to light and heat, have increasingly
replaced more substantial walls; but since 1974 there has been a reversal of
this trend as a result of the increasing cost of energy.

Walls can be designed as extended columns, in which case the rules of
column design discussed earlier in this chapter apply. Chapter 14 of the *ACI
Code* also admits an empirical design method, provided that the resultant
load falls within the middle third of the wall thickness.

The design axial load strength of a wall is

$$\phi P_{nw} = 0.55\phi f_c' A_g \left[1 - \left(\frac{k\ell_c}{32h} \right)^2 \right]$$

where the strength reduction factor $\phi = 0.70$, A_g is the gross cross-sectional
area of the wall, ℓ_c is the vertical distance between supports, k is a factor
that varies from 0.8 to 2.0 (depending on the degree of restraint), and h is
the overall wall thickness. Using this method, the minimum wall thickness
is 6 in. The minimum area of horizontal reinforcement is $0.0020A_g$, and the
minimum area of vertical reinforcement is $0.0012A_g$.

Shear walls are discussed in Section 11.6.

EXERCISES

1. What are the minimum and maximum ratios of the main reinforcement to the
   concrete in columns?
2. In beams and slabs it is customary to use a fairly large number of fairly small
   diameter bars. In columns it is customary to use relatively few bars of fairly large
   diameter. What is the reason for this difference?
3. Why is it necessary to tie the main reinforcing bars of columns? What is the
   minimum size of the ties and their maximum spacing?
4. How much concrete cover is needed for the reinforcement of columns for various
   bar sizes, indoors and outdoors?
5. Explain the difference between a circular column with tied reinforcement and one
   with spiral reinforcement. Why has a column with spiral reinforcement a higher
   load-bearing capacity for the same cross-sectional areas of concrete and of main
   reinforcement?
6. Why is the strength reduction factor for columns lower than for beams? Why is

there an extra strength reduction for concentrically loaded columns [(Eqs. (13.2) and (13.3)]?

7. What causes a column load to act eccentrically? What produces a bending moment in a column? How is a bending moment converted into an eccentricity of the column load?

8. What is biaxial bending? In which type of column is it particularly common?

9. Draw diagrams to show the movement of the neutral axis with increasing eccentricity, from outside the section (when all the main reinforcement is in compression) through a primary compression failure (as in a heavily reinforced beam, Section 9.3) to a primary tension failure.

10. Design a short, concentrically loaded column for the ground floor of a six-story building to carry a dead load of 250,000 lb and a live load of 100,000 lb per story, using Grade 5 concrete and Grade 60 main reinforcement.

11. Determine the load-bearing capacity of a 24-in.-square column of Grade 4 concrete, reinforced with 8 No. 14 bars of Grade 60 steel. Assume the column is concentrically loaded.

Metric Exercises

12. Design a short, concentrically loaded column for the ground floor of a six-story building to carry a dead load of 1000 kN and a live load of 500 kN per story, using Grade 35 concrete and Grade 400 main reinforcement.

13. Determine the load-bearing capacity of a 600-mm-square column of Grade 35 concrete, reinforced with 4 No. 55 bars of Grade 400 steel. Assume the column is concentrically loaded.

REFERENCES

13.1 *Ultimate Strength Design Handbook.* American Concrete Institute, Detroit, 1970. Two volumes.

13.2 *Metric Design Handbook for Reinforced Concrete Elements in Accordance with the Strength Design Method of CSA A23.3–M 1977,* Canadian Portland Cement Association, Ottawa, 1978. Looseleaf, about 800 pages.

Chapter 14

Footings and Retaining Walls

The design of foundations differs from other aspects of structural design because the behavior of the foundation soil is only imperfectly known. It is therefore a task for specialists, unless the soil has a reliable bearing capacity. For small buildings an adequate foundation can usually be constructed by assuming a sufficiently low bearing pressure for the soil.

This chapter deals briefly and descriptively with the subject. Only single-column and wall footings are discussed in more detail.

14.1 SQUARE COLUMN FOOTINGS

The distribution of soil pressure under a footing depends on the type of soil. However, by taking the bearing pressure sufficiently low, it is possible to assume a uniform distribution of pressure (Fig. 14.1).

The size of the footing is determined from the service loads, not the ultimate loads. We provide a footing that is just big enough to ensure that the safe bearing pressure under the column footing (Figs. 14.1 and 14.2) is not exceeded.

For the design of the cross section of the footing, however, we use the ultimate strength theory; that is, we multiply each of the service (dead, live, or wind) loads by its appropriate load factor from Eqs. (8.1) to (8.3). It is possible to obtain a different pattern of stress distribution from the factored

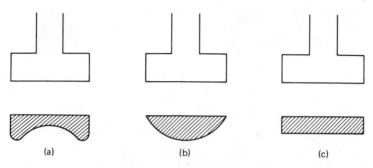

Figure 14.1 Distribution of soil pressure under a concentrically loaded footing.
(a) Elastic cohesive clay. (b) Sand or gravel. (c) Assumed distribution of pressure.

and from the unfactored loads.   For example, if the eccentricity of the column
load in Fig. 14.2 is mainly due to wind forces (which it frequently is), the
individual factoring of the loads produces a different pressure distribution.
In that case the *ACI Code Commentary* (Ref. 4.3) recommends that the
appropriate load factors be applied directly to the safe bearing pressures and
reactions obtained from the unfactored loading. Having ascertained the bearing pressures, which act like four systems of cantilever loads on the footing,
we can calculate the maximum shear force [Fig. 14.3(a)] and bending moment
[Fig. 14.3(b)].

  Shear more often determines the depth of the footing slab than does
bending moment or deflection. As in the case of flat slabs and flat plates
(Section 12.5), the *critical section for shear* is at a distance $\frac{1}{2}d$ from the column

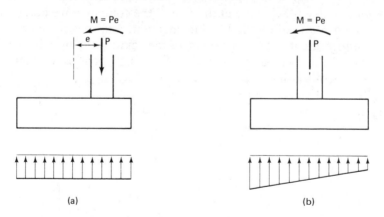

Figure 14.2 Column transmitting a bending moment to the foundation.   (a) If
the eccentricity of the column load is always in the same direction, we can achieve
uniform foundation pressure by balancing it with an eccentrically placed foundation
slab.   (b) If the bending moment is liable to reversal (e.g., if it is caused by wind
forces), a nonuniform foundation pressure cannot be avoided; it is then particularly
important to check that factoring of the working loads does not cause uplift of the
foundation slab.

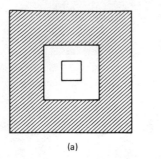
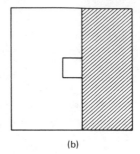

(a)   (b)

Figure 14.3 Critical sections in single-column footings. (a) For shear at a distance
½d from the column face, where d is the effective depth. The shaded portion is
assumed to be sheared relative to the unshaded portion of the footing. (b) For
bending at the face of the column. The shaded portion is bent as a cantilever.

face, where d is the effective depth.   The soil pressure acting upward on the
shaded area in Fig. 14.3(a) produces a shear force that is resisted by the shear
stress over the effective depth and the perimeter of the square

$$b_0 = 4(½d + b + ½d) = 4(b + d) \tag{14.1}$$

where b is the width of the column.

As in the case of flat plates and flat slabs, the shear stress in footings
without shear reinforcement is limited to $4\sqrt{f'_c}$.   If this stress is exceeded, it
is generally more economical to increase the thickness of the footing slab than
to use shear reinforcement.

The slab is also bent as a cantilever shown in Fig. 14.3(b).   The *critical
section for bending* is evidently at the face of the column.   We determine the
reinforcement required by the bending moment, and thus obtain a two-way
mat of reinforcement.   The anchorage of the reinforcement may be critical,
particularly if the foundation material has a high bearing pressure which
produces a small footing.

In heavily loaded cantilever footings, concrete can be saved by using a
sloping surface or steps (Fig. 14.4).   In both cases, but particularly in stepped
footings, it is necessary to check the shear strength and the anchorage at each
step, using the ½d appropriate to the step.

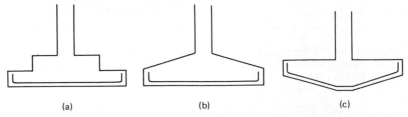

(a)   (b)   (c)

Figure 14.4 Stepped and sloped column footings.   (a) In a stepped footing the
shear must be checked at each step. (b) This sloped footing requires a dry concrete
mix.   (c) This sloped footing requires special excavation and bar bending.

14.2 FOOTING DESIGN EXAMPLE

Example 14.1A

Design a footing for a single column 20 in. square, carrying a dead load of 400,000 lb. and a live load of 160,000 lb. The concrete is Grade 4, the steel is Grade 60, and the permissible soil pressure is 3,000 psf.

Data given: $D = 400,000$ lb, $L = 160,000$ lb, $f'_c = 4,000$ psi, $f_y = 60,000$ psi, $\phi = 0.90$ for flexure and 0.85 for shear, $q_s = 3,000$ psf, and b (column) $= 20$ in.

The total service load is

$$W = 400,000 + 160,000 = 560,000 \text{ lb}$$

The total ultimate load is

$$U = 400,000 \times 1.4 + 160,000 \times 1.7 = 832,000 \text{ lb}$$

Width of footing: Let us assume a 30-in.-thick footing slab. It weighs $30 \times 12 = 360$ psf and this must be subtracted from the permissible soil pressure, because it constitutes a load additional to the column load. Hence the effective permissible soil pressure is

$$q = 3,000 - 360 = 2,640 \text{ psf}$$

The footing must therefore have a surface area of not less than

$$A = B^2 = \frac{W}{q} = \frac{560,000}{2,640} = 212.1 \text{ ft}^2$$

$$B = 14.56 \text{ ft.}$$

Use a footing 15 ft square.

Depth of footing for shear: The smaller effective depth of a 30-in. slab with reinforcement in both directions is

$$d = 30 - 3 \text{ (cover)} - 1 \text{ (assumed bar diameter)} - \tfrac{1}{2} \times 1 = 25.5 \text{ in.}$$

The shear force and bending moment are most conveniently obtained from the upward soil pressure at the ultimate load, which is the reverse of the procedure used in the design of the flat plate in Example 12.2. The ultimate soil pressure is

$$q_u = \frac{qU}{W} = \frac{2,640 \times 832,000}{560,000} = 3,922 \text{ psf}$$

The critical line for shear is at a distance $\tfrac{1}{2}d$ from the column face, that is, a square of

$$20 + 25.5 = 45.5 \text{ in.} = 3.79 \text{ ft.}$$

The ultimate shear force equals the pressure q_u acting upward on the area shaded in Fig. 14.5(a).

$$V_u = 3,922(15^2 - 3.79^2) = 826,000 \text{ lb}$$

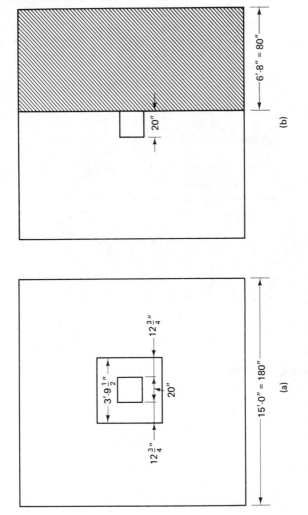

(a)

(b)

Figure 14.5 Plan of the footing of Example 14.1.A. (a) Upward soil pressure critical for shear. (b) Upward soil pressure critical for flexure.

and the nominal ultimate shear force is

$$V_n = \frac{V_u}{\phi} = \frac{826,000}{0.85} = 972,000 \text{ lb}$$

This acts on a circumference of

$$b_0 = 4 \times 45.5 = 182 \text{ in.}$$

The limiting shear stress in the slab of a single-column footing is $4\sqrt{f'_c} = 253$ psi. Therefore, the minimum effective depth for shear is

$$d = \frac{972,000}{182 \times 253} = 21.1 \text{ in. (25.5 in.)   O.K.}$$

Use a slab 30 in. thick.

   Flexural reinforcement: The bending moment is calculated for a cantilever spanning 6 ft 8 in. from the face of the column [Fig. 14.5(b)] loaded upward by the soil pressure.

$$M_u = \frac{1}{2} \times 3,922 \times 6.67^2 = 87,200 \text{ lb·ft/ft}$$

We will take z conservatively (Table 9.1) as $0.90d = 0.90 \times 25.5 = 23.0$ in.   The area of reinforcement required, from Eq. (9.8), is

$$A_s = \frac{87,200 \times 12}{0.9 \times 60,000 \times 23.0} = 0.843 \text{ in.}^2/\text{ft}$$

Use No. 7 bars of 8½-in. centers ($= 0.85$ in.²/ft); *22 bars are required in each direction.* All bars need hooks at both ends.
   Additional checks: The reinforcement ratio is

$$\rho = \frac{0.85}{12 \times 25.5} = 0.0028$$

so that the possibility that the concrete may be overstressed need not be considered. The minimum amount of reinforcement is

$$0.001\ 8 \times 12 \times 30 = 0.648 \text{ in.}^2/\text{ft (0.85 in.}^2/\text{ft)   O.K.}$$

The maximum spacing of the reinforcement is 3×30 or 18 in. (8½ in.)   O.K. The development length for a No. 7 bar, from Table 6.1, is 18.22 in.   O.K. The minimum overall depth for deflection of a cantilever, from Table 8.2, is

$$h \geqslant \frac{\ell}{20} = \frac{80}{20} = 4 \text{ in. (30 in.)   O.K.}$$

Example 14.1B

Design a footing for a single column 500 mm square, carrying a dead load of 1 600 kN and a live load of 800 kN. The concrete is Grade 25, the steel is Grade 400, and the permissible soil pressure is 150 kPa.
   Data given: $D = 1\ 600$ kN, $L = 800$ kN, $f'_c = 25$ MPa, $f_y = 400$ MPa, $\phi = 0.90$ for flexure and 0.85 for shear, $q_s = 150$ kPa, and b (column) $= 500$ mm.

The total service load is

$$W = 1600 + 800 = 2\,400 \text{ kN}$$

The total ultimate load is

$$U = 1\,600 \times 1.4 + 800 \times 1.7 = 3\,600 \text{ kN}$$

Width of footing: Let us assume an 800-mm-thick footing slab. It weighs $0.8 \times 24 = 19.2$ kPa, and this must be subtracted from the permissible soil pressure, because it constitutes a load additional to the column load. Hence the effective permissible soil pressure

$$q = 150 - 19.2 = 130.8 \text{ kPa}$$

The footing must therefore have a surface area of not less than

$$A = B^2 = \frac{W}{q} = \frac{2\,400}{130.8} = 18.35 \text{ m}^2$$

$$B = 4.28 \text{ m}$$

Use a footing 4.5 m square.

Depth of footing for shear: The smaller effective depth of an 800-mm slab with reinforcement in both directions is

$$d = 800 - 75 \text{ (cover)} - 25 \text{ (assumed bar diameter)} - \tfrac{1}{2} \times 25 = 687 \text{ mm}$$

The shear force and bending moment are most conveniently obtained from the upward soil pressure at the ultimate load, which is the reverse of the procedure used in the design of the flat plate in Example 12.2. The ultimate soil pressure is

$$q_u = \frac{qU}{W} = \frac{130.8 \times 3\,600}{2\,400} = 196.2 \text{ kPa}$$

The critical line for shear is at a distance $\tfrac{1}{2}d$ from the column face, that is, a square of

$$500 + 687 = 1\,187 \text{ mm}$$

The ultimate shear force equals the pressure q_u acting upward on the area shaded in Fig. 14.5(a).

$$V_u = 196.2(4.5^2 - 1.187^2) = 3.697 \text{ MN}$$

and the nominal ultimate shear force is

$$V_n = \frac{V_u}{\phi} = \frac{3.697}{0.85} = 4.349 \text{ MN}$$

This acts on a circumference

$$b_0 = 4 \times 1\,187 = 4.748 \text{ m}$$

The limiting shear stress in the slab of a single-column footing is $\tfrac{1}{3}\sqrt{f_c'} = 1.67$ MPa. Therefore, the minimum effective depth for shear is

$$d = \frac{4.349}{4.748 \times 1.67} = 0.548 \text{ m } (0.687 \text{ m}) \quad \text{O.K.}$$

We only need an effective depth of 548 mm and we have 687 mm; however, if we reduce the overall thickness of the slab to 700 mm, we do not have sufficient depth for shear. q and q_u are increased slightly.   The footing remains 4.5 m square.   The ultimate shear force becomes

$$V_u = 200(4.5^2 - 1.087^2) = 3.814 \text{ kN}$$

$$b_0 = 4\ 346 \text{ mm}$$

and

$$d = \frac{3.814}{0.85 \times 4.346 \times 1.67} = 0.618 \text{ m}$$

which is not available.   *Use a slab 800 mm thick.*

Flexural reinforcement: The bending moment is calculated for a cantilever spanning 2.000 m from the face of the column, loaded upward by the soil pressure.

$$M_u = \tfrac{1}{2} \times 196.2 \times 10^{-3} \times 2^2 = 0.392 \text{ MN·m per meter width of slab}$$

We will take z conservatively (Table 9.2) as $0.90d = 0.90 \times 687 = 618$ mm.   The area of reinforcement required, from Eq. (9.8), is

$$A_s = \frac{0.392}{0.9 \times 400 \times 0.618} = 1.762 \times 10^{-3} \text{ m}^2/\text{m} = 1\ 748 \text{ mm}^2/\text{m}$$

Use No. 25 bars on 250-mm centers ($= 2000$ mm²/m); *18 bars are required in each direction.*   All bars need hooks at both ends.

Additional checks: The reinforcement ratio

$$\rho = \frac{2\ 000}{1\ 000 \times 687} = 0.0029$$

so that the possibility that the concrete may be overstressed need not be considered.

The minimum amount of reinforcement is

$$0.001\ 8 \times 1\ 000 \times 800 = 1\ 440 \text{ mm}^2/\text{m} \ (2\ 000 \text{ mm}^2/\text{m}) \quad \text{O.K.}$$

The maximum spacing of the reinforcement is 3×800 or 450 mm (200 mm)   O.K.
The development length for a 25-mm bar, from Table 6.1, is 608 mm.   O.K.
The minimum overall depth for deflection of a cantilever, from Table 8.2, is

$$h \geqslant \frac{\ell}{20} = \frac{2\ 000}{20} = 100 \text{ mm} \ (800 \text{ mm}) \quad \text{O.K.}$$

14.3 COMBINED FOOTINGS

If a column is so close to the boundary of the property or to another building that there is insufficient room for the cantilevers, we must combine its footing with that of the adjacent column to form a beam (Fig. 14.6).

If the bearing pressures are low or the column loads are very high, we may not have enough room for individual column footings, and they must be

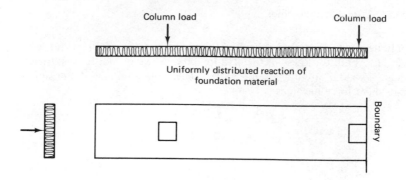

Figure 14.6 Combined two-column footings are required when one column is hard up against a boundary or another building, and the footing cannot be extended beyond its face.

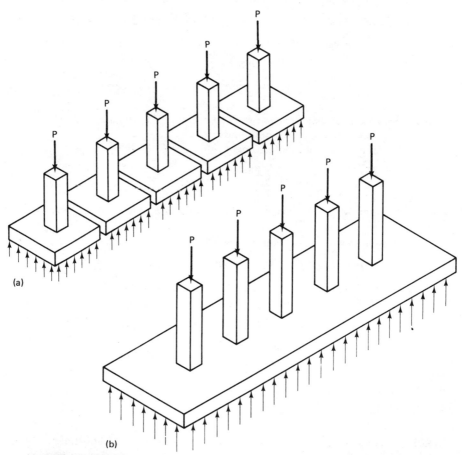

Figure 14.7 (a) Individual column footings behave like cantilevered slabs. (b) A strip footing behaves like a continuous slab. When the cantilevered footings get close to one another, there are evident advantages in merging the individual column footings into strip footings.

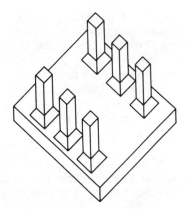

Figure 14.8 When the strip footings come close to one another, it is advantageous to combine them into a single raft foundation, which behaves like a flat slab turned upside down.

merged into strips that behave like continuous slabs or beams (Fig. 14.7). As buildings get taller or foundation materials weaker (or both), we must resort to a raft foundation, which is essentially a flat slab (see Section 12.5) turned upside down (Fig. 14.8).

Example 14.2

The footing of the columns in each of the five rows of the building shown in Fig. 14.9(a) are to be combined into strip footings not more than 3 m wide.   Determine the highest number of stories for which these strip footings are suitable, if each interior

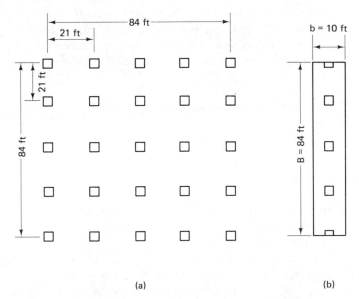

(a)   (b)

Figure 14.9 Example 14.2. (a) Plan of the columns for a multistory building. (b) Layout of the strip footings. When the strip footings become wider than 10 ft, the gap between them is reduced to less than 11 ft, and it becomes more economical to employ a raft foundation.

column carries a load of 70,000 lb per floor, and each exterior column carries half that load. The maximum permissible foundation pressure is 5,000 psf.

Data given: $P = D + L = 70,000$ lb, $q_s = 5,000$ psi.

Each floor contributes $(3 + 2 \times \frac{1}{2})P$, so that the total load transmitted to the foundation by the columns for n floors is $4nP$. The strip footings are 84 ft long, and are to be no more than 10 ft wide. Therefore,

$$5,000 \times 84 \times 10 = 4 \times n \times 70,000$$

and $n = 15.0$.

When the building exceeds 15 stories in height, it would be more economical to use a raft foundation, which because of its continuity between the strips has lower bending moments and therefore a smaller thickness of concrete.

14.4 RETAINING WALLS

The principal types of retaining wall are shown in Fig. 14.10. When there is ample room and easy access, a gravity retaining wall (a) of plain concrete or masonry has some advantages; it need only be designed against overturning. The standard cantilever retaining wall is shown in (b). The weight of the soil pressing on the heel provides the resistance against sliding, while the toe gives the wall a greater leverage against overturning.

The pressure behind the wall builds up linearly with the depth of the wall (Fig. 14.11) and this produces a horizontal force H pressing against the wall. The combined effect of this horizontal force, and of the vertical force W due to the weight of the wall and the soil resting on the heel, produces a linearly varying foundation pressure P. The wall must be so proportioned that:

1. The maximum foundation pressure under the toe is less than that permissible.
2. There is pressure under the heel, not uplift.
3. The frictional resistance μP is greater than the horizontal force pushing the wall outward.

Having determined that the wall as a whole is safe, we must determine the bending moments in the three cantilevers: the wall, the heel, and the toe. The load on each is shown in Fig. 14.12.

From these we can calculate the thickness of the slabs and the amount of reinforcement needed. The reinforcement in the wall slab can be reduced toward the top as the bending moment decreases, by increasing the spacing. At the bottom end the reinforcement is spliced to the bars in the toe which project into the slab, to transfer the bending moment to the base [Fig. 14.12(c)]. The

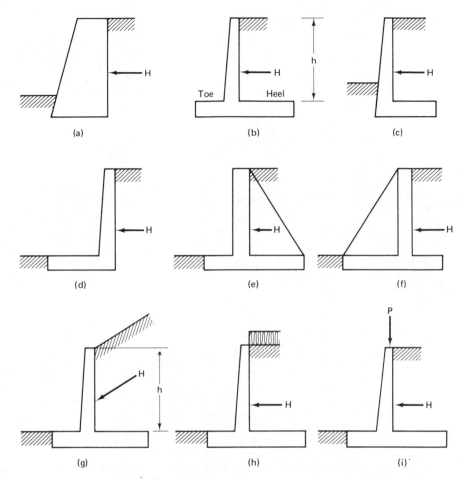

Figure 14.10   Retaining walls.   (a) Gravity wall.   (b) Standard cantilever wall. (c) Cantilever wall without toe.   (d) Cantilever wall without heel.   (e) Counterfort wall.   (f) Buttress wall.   (g) Cantilever wall with inclined back fill.   (h) Cantilever wall with surcharge.   (i) Cantilever wall with vertical load.

reinforcement in the heel is on the opposite side of the slab and therefore quite separate.   The anchorage of the steel in both the heel and the toe at the free ends requires attention.   Shear reinforcement is rarely used: if the shear stresses are too high, it is cheaper to use a thicker slab.

The base slab must be cast first, and between it and the vertical wall there must be a construction joint with a key capable of resisting the horizontal shear.

Let us now look briefly at other types of retaining wall.   When the wall comes up against a property boundary or against an existing building it may be necessary to build a cantilever wall without a toe [Fig. 14.10(c)] or without

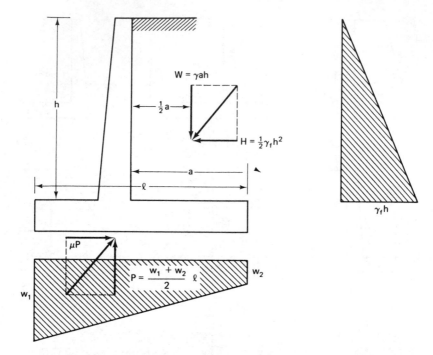

Figure 14.11 Forces acting on a cantilever retaining wall. The pressure builds up linearly behind the wall to $\gamma_f h$, where γ_f is the equivalent fluid weight of the backing soil. This produces a resultant force $H = \frac{1}{2}\gamma_f h^2$. The weight of the solid resting on the heel $W = \gamma ah$, where γ is the unit weight of the backing soil. This produces a pressure P under the foundation, which in turn produces a frictional force between the solid and the concrete μP, where μ is the coefficient of friction; μP must at least equal H. In addition, the resultant of W and H must line up with the resultant P and μP. In practice the problem is a little more complicated because we must include the quite appreciable weight of the concrete retaining wall itself.

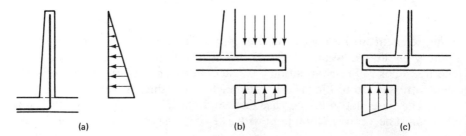

Figure 14.12 Forces acting on the three cantilevers of the standard wall, and the placing of the reinforcement. (a) The wall is a cantilever resisting the linearly increasing horizontal backing pressure. (b) The heel is a cantilever resisting the vertical backing pressure, reduced by the upward foundation pressure. (c) The toe is a cantilever resisting the upward foundation pressure.

a heel (d).   When the wall becomes too tall, we use counterforts (e) or buttresses (f). The wall then spans as a continuous horizontal slab between the counterforts or buttresses, and it is the latter that form the cantilevers. Counterforts offer no obstruction and the wall looks neater; they also act as T-beams, with the slab of the retaining wall as the compression flange.   However, they need to be tied to the wall with stirrups, whereas buttresses press against the wall.   A wall retaining soil may have a surcharge (g), and a wall forming part of a building may carry a load on top of the soil retained (h) or act as a load-bearing wall (i).

EXERCISES

1. Under what circumstances would you use a single-column footing, a strip footing, and a raft foundation? Consider both the height of the building and the strength of the foundation material.

2. How would you obtain approximately uniform soil pressure under a single-column footing for an eccentrically loaded column?

3. How would you obtain approximately uniform soil pressure under a combined column footing for two columns, of which one is immediately adjacent to the property boundary.

4. What is meant by the heel and the toe of a reinforced concrete cantilever retaining wall? How does the weight of the soil resting on the heel affect the stability of the wall?

5. What is meant by a buttress and by a counterfort in relation to a retaining wall? How does each affect the structural function of (a) the wall slab and (b) the floor slab?

6. A column, 24 in. square, is supported on a 14-ft-square slab footing. The upward soil pressure at the ultimate load is 4,500 psf. Determine the effective depth of the footing slab if the grade of the concrete is 3.5.

7. Determine the amount of Grade 60 reinforcement required for the footing slab of Exercise 6 and the thickness of the slab.

Chapter 15

Prestressed Concrete

Prestressed concrete is a structural material in its own right, with a technology as complex as that of reinforced concrete. To explain it fully would require a separate book. This chapter explains briefly the principles.

15.1 BRIEF HISTORY OF PRESTRESSED CONCRETE

A concept of prestressing has been used for many centuries. The traditional wooden barrel consists of wooden staves on which iron hoops are hammered. As the barrel is filled with liquid, the hoop pressure expands the barrel, and it would leak if the staves had not been prestressed by the iron hoops. If the liquid pressure exceeds the prestress, gaps open between the staves, and the barrel starts to leak.

From the eighteenth to the early twentieth century heavy guns were made in two or more parts. The bore of the outer part was made a little smaller than the outer diameter of the inner part. The outer part was then heated, and in this expanded condition was slid over the inner part. As it cooled it prestressed the inner barrel. The gun could not burst until the prestress had been overcome.

The advantages of prestressing a material that cracks as easily as concrete are evident (see Section 1.3), and the first attempts were made by C. W. D. Doehring in Germany in 1886 and by P. H. Jackson in the United States in

1888.   This was only a little over 30 years after the first attempts to reinforce concrete.   However, whereas reinforced concrete was immediately success-ful, nineteenth-century attempts to prestress concrete failed (Ref. 5.1, p. 59). The prestress disappeared mysteriously after a few weeks.   The problem of loss of prestress was solved only in 1921 by Karl Wettstein in Austria, who used high-tensile wire.   Post-tensioned concrete (see Section 15.2) was in-vented by E. Freyssinet in France in 1930.

Prestressed concrete uses far less steel than reinforced concrete (see Section 15.4), and it was therefore widely used during the 1939–1945 war in Germany and German-occupied Europe for military installations.

After 1945 prestressed concrete came into general use, and it was hailed as the successor to reinforced concrete because the new material did not have cracks.   But it was also more expensive, and when the postwar steel shortage ended, prestressed concrete became less attractive.

Prestressed concrete is an excellent material for highway bridges.   Steel bridges need periodical repainting, but prestressed concrete is virtually main-tenance free.   The spans are comparatively large when one highway crosses another at a different level, and prestressed concrete girders are then more economical and more elegant.   The fact that a number of identical girders are required is also more advantageous for prestressed concrete than for cast-in-place reinforced concrete.

In architecture, prestressed concrete has not played the role predicted for it in the late 1940s.   However, it is useful when spans are comparatively large.   It is possible to use a few prestressed beams or slabs in a building which is otherwise constructed of normal reinforced concrete.

15.2 PRETENSIONING AND POSTTENSIONING

There are three types of prestressed concrete.   Concrete can be prestressed against *fixed anchorages* (Fig. 15.1) and only nominal reinforcement is then required.   One of the most notable examples is the Gladesville Bridge in Sydney, at 1,000 ft (305 m) still the world's longest-spanning concrete arch (Fig. 3.13), which has only nominal reinforcement.   However, few buildings provide rigid anchorages.

Alternatively, steel can be tensioned against a strong mold, and the

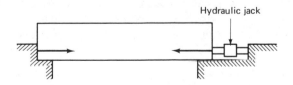

Figure 15.1   Prestressing against fixed anchorage.

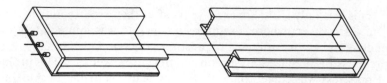

Figure 15.2 Pretensioning.

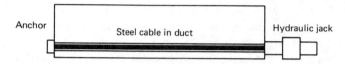

Figure 15.3 Posttensioning.

concrete then cast around it (Fig. 15.2). When the concrete has gained sufficient strength the wires are released, and the concrete is compressed. This is called *pretensioning*, because the steel is tensioned before the concrete is cast. The wires and strands used for pretensioning are greatly reduced in diameter by the high prestress (see Section 15.3), and when released they regain their original diameter at the free ends. The wedge so formed aids the anchorage provided by bond between the steel and the concrete.

Pretensioning can only be done in a factory, and pretensioned beams are therefore always precast. They are generally designed to be simply supported.

In *posttensioned* units the concrete is cast around steel bars, wires, or strands in ducts that permit their movement. After the concrete has hardened sufficiently, the steel is tensioned by a hydraulic jack (Fig. 15.3), which uses the end of the concrete beam or slab as its anchorage. The prestress is then anchored to the ends of the concrete by nuts on screwed bars, or by wedging individual wires or strands in slots or holes.

The ducts are normally grouted after the prestressing operation is complete. This protects the steel against corrosion, and it establishes contact between the steel and the surrounding concrete. The strength of posttensioned concrete under the service loads is not greatly affected by grouting, but the ultimate strength of nongrouted beams is much lower than that of grouted beams (see Fig. 15.5).

Posttensioning can be done in a factory, but it can also be done on cast-in-place concrete. Thus posttensioning can be used for a few long spans in a cast-in-place reinforced concrete structure, or for long-span flat plates (see Sections 12.5 and 15.7) in an otherwise normally reinforced concrete frame.

15.3 LOSS OF PRESTRESS

Concrete shrinks after it has been cast. The contraction is approximately 3 $\times$ 10^{-4} in./in. (Section 8.2). Although this appears very small, it can cause an appreciable loss of prestress because the elastic extension is also very small. The modulus of elasticity of steel is 29,000,000 psi, so that a contraction of 3×10^{-4} produces a loss of prestress of

$$29,000,000 \times 3 \times 10^{-4} = 8700 \text{ psi}$$

If Grade 40 steel with a maximum permissible stress of 20,000 psi (Section 8.1) were used, the loss of prestress due to shrinkage alone would be 43%. Most of the rest would be lost due to creep because the prestress is a sustained load that continues to squeeze water from the concrete. The early experiments used a normal reinforcing steel, and that is why they failed.

In 1921, Karl Wettstein used high-tensile wire that could be prestressed to 200,000 psi, and the loss due to shrinkage was thereby reduced to 8700/200,000, or 4%, which is acceptable.

Apart from losses due to shrinkage and creep of the concrete, there are losses due to the stress relaxation of the steel at sustained high stresses, and due to slipping of the steel at the anchorage when friction grips are used. In pretensioned concrete the release of the steel from the anchorage of the mold or casting bed causes an elastic shortening of the concrete that results in a loss of prestress.

In posttensioned concrete, cables are generally curved (see Section 15.7), and this produces a loss of prestress at the center of the beam due to friction between the steel and the duct in which it is prestressed (Fig. 15.3). The friction loss can be high if the curvature of the cables is too great. This can be visualized by noting that a ship can be firmly held by a rope or cable wrapped 1½ turns around a bollard.

The following causes of loss of prestress should be considered for pretensioned concrete: shrinkage of concrete, creep of concrete, stress relaxation of steel, and elastic shortening of concrete; and for posttensioned concrete: shrinkage of concrete (a much lower loss), creep of concrete, stress relaxation of steel, slipping at the anchorages, and friction in curved ducts. These losses can be calculated, and this is explained in most books on prestressed concrete (e.g., Ref. 15.2), and it is discussed in more detail by the Prestressed Concrete Institute (Ref. 15.3). For simple calculations it is sufficient to take the loss of prestress as 36,000 psi (250 MPa) for wires and strands.

Owing to loss of prestress, the steel never regains the stress it had during the initial prestressing operation. The steel stress increases as the load is placed on a prestressed beam or slab, but the increase is less than the loss of prestress. Steel stresses need not therefore be checked, except for the prestressing operation.

15.4 MATERIALS USED FOR PRESTRESSED CONCRETE

Because of the loss of prestress, which is to a large extent independent of the strength of the steel, only high-tensile steel can be used economically in prestressed concrete. It is customary to specify the ultimate because high-

Figure 15.4 Seven-wire strand.

tensile steel has no pronounced yield stress.

The ultimate steel stress f_u for high-tensile deformed bars ⅝ to 1½ in. in diameter is about 150,000 psi (1 050 MPa). The ultimate stress of 0.2 to 0.275 in. (5-mm and 7-mm) diameter wires and of seven-wire strands (Fig. 15.4) made from these wires ranges from 220,000 to 280,000 psi (1 500 to 1 900 MPa).

The generic term for all types of prestressing steel is *tendon*, which may mean a high-tensile bar, a wire, a strand, or a cable.

It would not be economical to use low-strength concrete with high-tensile steel, and concrete is usually Grade 5 to Grade 7.

In reinforced concrete the steel stress is limited to ensure that the cracks do not get too large. The *ACI Code* (see Section 6.2) limits the yield stress of reinforcing steel to 80,000 psi. The most commonly used reinforcing steel has a yield stress f_y = 60,000 psi and an ultimate stress f_u = 90,000 psi. The steel used in prestressed concrete is thus two to three times as strong.

The ultimate strength is not, however, altered by the fact that a beam is prestressed. It is still necessary to provide a resistance moment consisting of a tensile force and a moment arm (Figs. 9.2 and 15.5).

$$M_u = Tz$$

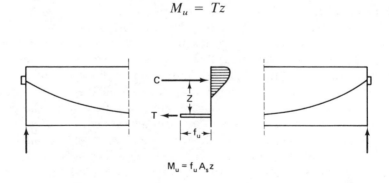

$$M_u = f_u A_s z$$

Figure 15.5 Stress distribution in a posttensioned concrete beam at the ultimate load, immediately prior to failure. The steel stress is f_u, irrespective of the magnitude of the prestress, because in a tension failure steel must be stressed to the ultimate stress. If the ducts are not grouted, the ultimate strength is reduced, because the steel stress is independent of the stress in the surrounding concrete. The steel stress equalizes along the entire length of the tendon, and the overall extension of the tendon is consequently much greater. As a result, the ultimate steel stress f_u is not reached before the concrete at midspan is crushed.

The moment arm z at the ultimate load is approximately the same for both reinforced and prestressed concrete beams of the same depth.   The tensile force is

$$f_s A_s$$

for both.   Thus if we use steel three times as strong, we can reduce the amount of steel to one-third.

15.5 PRINCIPLES OF PRESTRESSED CONCRETE DESIGN

The rectangular section shown in Fig. 15.6 has a concentric tendon tensioned to a prestress P.   This applies a compressive stress to the concrete

$$f_c = \frac{P}{A}$$

where A is the cross-sectional area of the concrete.

If we now place a load on a beam which produces a bending moment M, we add compression on top and tension on the bottom.   If tension in the concrete is not permitted, the maximum permissible bending moment is one that produces a tensile stress just canceling the compression due to prestress:

$$f_t = \frac{M}{S} \le f_c = \frac{P}{A}$$

where S is the section modulus.

We can take two steps to improve the performance of this beam.   One is to use an I-section, which for the same weight and cross-sectional area A gives a higher section modulus S.   This is always done in steel structures, but rarely in reinforced concrete structures, because the extra labor does not compensate for the saving in material.   In the more expensive prestressed concrete, I-sections or T-sections are often worthwhile.

Second, we can place the prestressing tendon eccentrically, so that it superimposes a flexural stress Pe/S on the compressive stress P/A.   This produces tension on top (which is immediately canceled out by the bending moment due to the weight of the beam), and compression on the bottom (Fig. 15.7).   We can thus place a much higher useful load on the beam.   This

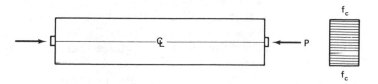

Figure 15.6 Concentric prestressing of rectangular section.

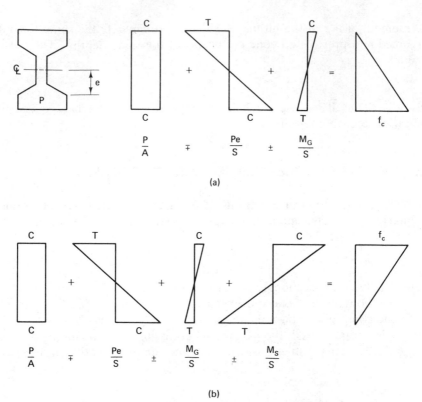

Figure 15.7   Eccentric prestressing of an I-section.   (a) During prestressing when the beam is subject to the prestress P at an eccentricity e and a bending moment M_G due to the weight of the beam itself.   (b) Under full load, when it is also subject to a bending moment M_S due to a superimposed dead and live load (C = compression; T = tension).   The aim is to produce zero stress on top and the full permissible compressive stress f_c at the bottom of the section during prestressing, and to reverse the condition under full load.

is the normal method for designing prestressed concrete beams.   It considers their behavior under the action of the service (i.e., the actual) loads.

We noted in Section 8.1 that there are two methods of designing reinforced concrete: the now obsolescent elastic method, which considers the behavior under the action of the service loads, and the ultimate-strength method, which deals with the determination of the factored load at which failure occurs.   These are alternative design methods: we can use one or the other, and either method gives a safe design.   Generally, the ultimate-strength method gives slightly smaller dimensions for reinforced concrete sections, so that it is the more economical one to use.

In prestressed concrete design there are also two design methods; one considers the behavior of prestressed concrete under the service loads (Fig.

15.7), and the other considers the ultimate strength (Fig. 15.5). These are not, however, alternative design conditions; both must be checked. Prestressed concrete is designed to be free from cracks under the (actual) service load, but before the ultimate strength is reached, the concrete has cracked under the (factored) ultimate load. These two conditions are so dissimilar that a safe design for the service load does not necessarily ensure that there is an adequate load factor against failure. In this brief discussion ultimate loads will not be considered, because they are only occasionally critical for design.

As for normal reinforced concrete design, shear must be checked, but the compressive forces due to the prestress reduce the diagonal tension due to shear.

15.6 DESIGN OF PRESTRESSED CONCRETE BEAMS

Example 15.1A

An I-section, with the dimensions shown in Fig. 15.8A, is post tensioned with 2.8 in.^{2} of tendons grouted after prestressing, placed at an average eccentricity of 10 in. The beam is simply supported over a span of 41 ft. The beam carries a super-imposed load of 3,500 lb/ft in addition to its own weight. Determine the stresses immediately after posttensioning, and under the action of the superimposed load. The ultimate steel stress is 270,000 psi, and the concrete is Grade 6.

Data given: Cross-sectional dimensions in Fig. 15.8.A, $e = 10$ in., $f_u = 270,000$ psi, $f'_c = 6,000$ psi, $A_s = 2.8$ in.^{2}, $\ell = 41$ ft, and $w_s = 3500$ lb/ft.

Properties of the concrete section: The cross-sectional area is

$$A = 18 \times 36 - 12 \times 24 = 360 \text{ in.}^2 = 2.5 \text{ ft}^2$$

The moment of inertia is

$$I = \frac{1}{12} 18 \times 36^3 - 12 \times 24^3 = 56,160 \text{ in.}^2$$

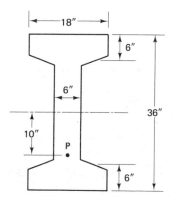

Figure 15.8A Dimensions of section analyzed in Examples 15.1A and 15.2A.

The section modulus is

$$S = \frac{I}{\frac{1}{2} \times 36} = 3{,}120 \text{ in.}^3$$

Bending moments: Concrete weighs 144 lb/ft³, and the prestressed concrete beam weighs 2.5 × 144 = 360 lb/ft. The bending moment due to the weight of the beam is

$$M_G = \tfrac{1}{8} \times 360 \times 41^2 = 75{,}645 \text{ lb·ft} = 908{,}000 \text{ lb·in.}$$

The bending moment due to the superimposed load is

$$M_S = \tfrac{1}{8} \times 3500 \times 41^2 \times 12 = 8{,}825{,}000 \text{ lb·in.}$$

Prestressing forces: The *ACI Code*, Section 18.5, limits the steel stress to

$$0.7f_u = 0.7 \times 270{,}000 = 189{,}000 \text{ psi}$$

We will assume a loss of prestress of 36,000 psi, so that the steel stress when the load is superimposed is 189,000 − 36,000 = 153,000 psi. Therefore, the initial prestress is

$$P_i = 189{,}000 \times 2.8 = 529{,}200 \text{ lb}$$

and the prestress remaining after losses is

$$P_r = 153{,}000 \times 2.8 = 428{,}400 \text{ lb}$$

Maximum permissible concrete stresses: Assuming that prestressing is delayed until the concrete reaches the specified strength of 6000 psi, the maximum permissible concrete stresses during the prestressing operation (*ACI Code*, Section 18.4) are

$$\text{In compression:} \quad 0.60f'_c = 3600 \text{ psi}$$

$$\text{In tension:} \quad 3\sqrt{f'_c} = 232 \text{ psi}$$

Under the action of the superimposed load,

$$\text{In compression:} \quad 0.45f'_c = 2700 \text{ psi}$$

$$\text{In tension:} \quad 6\sqrt{f'_c} = 465 \text{ psi}$$

Calculated concrete stresses: Initial concrete stresses at the top of the beam are

$$f_{cTi} = \frac{P_i}{A} - \frac{P_i e}{S} + \frac{M_G}{S}$$

$$= \frac{529{,}200}{360} - \frac{529{,}200 \times 10}{3120} + \frac{908{,}000}{3120}$$

$$= 1470 - 1696 + 291 = 65 \text{ psi compression} \quad (232 \text{ psi tension permitted})$$

At the bottom of the beam,

f_{cBi} = 1470 + 1696 − 291 = 2875 psi compression (3600 psi permitted)

Concrete stresses with full superimposed load at the top of the beam are

$$f_{cTr} = \frac{P_r}{A} - \frac{P_r e}{S} + \frac{M_G}{S} + \frac{M_S}{S}$$

$$= \frac{428,400}{360} + \frac{428,400 \times 10}{3120} + \frac{908,000 + 8,825,000}{3120}$$

$$= 1190 - 1373 + 3120 = 2937 \text{ psi compression (2700 psi permitted)}$$

At the bottom of the beam,

f_{cBr} = 1190 + 1373 − 3120 = 557 psi tension   (465 psi permitted)

Example 15.1B

An I-section, with the dimensions shown in Fig. 15.8B, is posttensioned with 1 820 mm^2 of tendons grouted after prestressing, placed at an average eccentricity of 250 mm.   The beam is simply supported over a span of 12.5 m.   The beam carries a superimposed load of 50 kN/m in addition to its own weight.   Determine the stresses immediately after posttensioning, and under the action of the superimposed load. The ultimate steel stress is 1 860 MPa, and the concrete is Grade 40.

Data given: Cross-sectional dimensions in Fig. 15.8, e = 250 mm, f_u = 1 860 MPa, f'_c = 40 MPa, A_s = 1 820 mm^2, ℓ = 12.5 m, and w_s = 50 kN/m.

Properties of the concrete section: The cross-sectional area is

$$A = 0.45 \times 0.9 - 0.3 \times 0.6 = 0.225 \text{ m}^2$$

The moment of inertia is

$$I = \frac{1}{12} (0.45 \times 0.9^3 - 0.3 \times 0.6^3) = 0.021 \ 94 \text{ m}^4$$

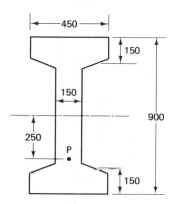

Figure 15.8B   Dimensions of section analyzed in Examples 15.1B and 15.2B.

The section modulus is

$$S = \frac{I}{\frac{1}{2} \times 0.9} = 0.048\ 8 \text{ m}^3$$

Bending moments: Concrete weighs 24 kN/m³, and the prestressed concrete beam weighs $0.225 \times 24 = 5.4$ kN/m. The bending moment due to the weight of the beam is

$$M_G = \tfrac{1}{8} \times 5.4 \times 10^{-3} \times 12.5^2 = 0.105\ 5 \text{ MN·m}$$

The bending moment due to the superimposed load is

$$M_S = \tfrac{1}{8} \times 50 \times 10^{-3} \times 12.5^2 = 0.977 \text{ MN·m}$$

Prestressing forces: The *ACI Code*, Section 18.5, limits the steel stress to

$$0.7f_u = 0.7 \times 1\ 860 = 1\ 302 \text{ MPa}$$

We will assume a loss of prestress of 250 MPa, so that the steel stress when the load is superimposed is $1\ 302 - 250 = 1\ 052$ MPa. Therefore, the initial prestress is

$$P_i = 1\ 302 \times 1.82 \times 10^{-3} = 2.370 \text{ MN}$$

and the prestress remaining after losses is

$$P_r = 1\ 052 \times 1.82 \times 10^{-3} = 1.915 \text{ MN}$$

Maximum permissible concrete stresses: Assuming that prestressing is delayed until the concrete reaches the specified strength of 40 MPa, the maximum permissible concrete stresses during the prestressing operation (*ACI Code*, Section 18.4) are

$$\text{In compression: } 0.60f'_c = 24 \text{ MPa}$$

$$\text{In tension: } 0.25\sqrt{f'_c} = 1.6 \text{ MPa}$$

Under the action of the superimposed load,

$$\text{In compression: } 0.45f'_c = 18 \text{ MPa}$$

$$\text{In tension: } 0.50\sqrt{f'_c} = 3.2 \text{ MPa}$$

Calculated concrete stresses: Initial concrete stresses at the top of the beam are

$$f_{cTi} = \frac{P_i}{A} - \frac{P_i e}{S} + \frac{M_G}{S}$$

$$= \frac{2.370}{0.225} - \frac{2.370 \times 0.250}{0.048\ 8} + \frac{0.105\ 5}{0.048\ 8}$$

$$= 10.533 - 12.141 + 2.162 = 0.55 \text{ MPa compression}$$
$$(1.6\text{-MPa tension permitted})$$

At the bottom of the beam,

$$f_{cBi} = 10.533 + 12.141 - 2.162 = 20.51 \text{ MPa compression} \quad (24 \text{ MPa permitted})$$

Concrete stresses with full superimposed load at the top of the beam are

$$f_{cTr} = \frac{P_r}{A} - \frac{P_r e}{S} + \frac{M_G}{S} + \frac{M_S}{S}$$

$$= \frac{1.915}{0.225} - \frac{1.915 \times 0.250}{0.048\ 8} + \frac{0.105\ 5 + 0.977}{0.048\ 8}$$

$$= 8.511 - 9.810 + 22.182 = 20.88 \text{ MPa compression}\quad (18 \text{ MPa permitted})$$

At the bottom of the beam,

$$f_{cBr} = 8.511 + 9.810 - 22.182 = 3.86 \text{ MPa tension}\quad (3.2 \text{ MPa permitted})$$

Example 15.2A

An I-section, with the dimensions shown in Fig. 15.8A, is posttensioned with 2.8 in.^{2} of tendons grouted after prestressing, placed at an average eccentricity of 10 in.   The ultimate steel stress is 270,000 psi, and the concrete is Grade 6.   The beam is simply supported over a span of 41 ft.   Determine the maximum permissible superimposed load.

Data given: Cross-sectional dimensions in Fig. 15.8A, $e = 10$ in., $A_s = 2.8$ in.^{2}, $l = 41$ ft $= 492$ in., $f_u = 270,000$ psi, and $f'_c = 6000$ psi.

The *maximum permissible concrete stresses* are the same as in Example 15.1A.

Properties of the concrete section: As calculated in Example 15.1A,

$$A = 360 \text{ in.}^2 \quad \text{and} \quad S = 3120 \text{ in.}^3$$

Bending moments: As calculated in Example 15.1A,

$$M_G = {'}908,000 \text{ lb·in.}$$

$$M_S = \tfrac{1}{8} \times w_s \times 492^2 \text{ lb·in.}$$

Prestressing forces: As calculated in Example 15.1A,

$$P_i = 529,200 \text{ lb} \quad \text{and} \quad P_r = 428,400 \text{ psi}$$

Concrete stresses: Initial concrete stresses, as calculated in Example 15.1A, are

$$f_{cTi} = 65 \text{ psi compression}\quad \text{O.K.}\quad (232 \text{ psi tension admissible})$$

$$f_{cBi} = 2875 \text{ psi compression}\quad \text{O.K.}\quad (3600 \text{ psi compression admissible})$$

Concrete stresses with full superimposed load at the top of the beam are

$$f_{cTr} = \frac{P_r}{A} - \frac{P_r e}{S} + \frac{M_G}{S} + \frac{M_S}{S} = 2700 \text{ psi}\quad \text{(permissible compressive stress)}$$

$$= \frac{428,400}{360} - \frac{428,400 \times 10}{3120} + \frac{908,000}{3120} + \frac{M_S}{S} = 2700$$

$$= 1190 - 1373 + 291 + \frac{M_S}{S} = 2700$$

$$\frac{M_S}{S} = 2592 \text{ psi}$$

At the bottom of the beam,

$$f_{cBr} = 1190 + 1373 - 291 - \frac{M_S}{S} = -465 \quad \text{(permissible tensile stress)}$$

$$\frac{M_S}{S} = 2737 \text{ psi}$$

The stress at the top of the beam is critical:

$$M_S = 2592 \times 3120$$

$$= \frac{1}{8} \times w_S \times 492^2$$

and the maximum permissible superimposed load is

$$w_S = 267.3 \text{ lb/in.} = 3207 \text{ lb/ft}$$

Example 15.2B

An I-section, with the dimensions shown in Fig. 15.8B, is posttensioned with 1 820 mm² of tendons grouted after prestressing, placed at an average eccentricity of 250 mm. The ultimate steel stress is 1 860 MPa, and the concrete is Grade 40. The beam is simply supported over a span of 12.5 m. Determine the maximum permissible superimposed load.

Data given: Cross-sectional dimensions in Fig. 15.8B, $e = 250$ mm, $A_s = 1\ 820$ mm², $\ell = 12.5$ m, $f_u = 1\ 860$ MPa, and $f_c' = 40$ MPa.

The *maximum permissible concrete stresses* are the same as in Example 15.1 B.

Properties of the concrete section: As calculated in Example 15.1B,

$$A = 0.225 \text{ m}^2 \quad \text{and} \quad S = 0.048\ 8 \text{ m}^3$$

Bending moments: As calculated in Example 15.1B,

$$M_G = 0.105\ 5 \text{ MN·m}$$

$$M_S = \frac{1}{8} \times w_s \times 10^{-3} \times 12.5^2 \text{ MN·m}$$

Prestressing forces: As calculated in Example 15.1B,

$$P_i = 2.370 \text{ MN} \quad \text{and} \quad P_r = 1.915 \text{ MN}$$

Concrete stresses: Initial concrete stresses, as calculated in Example 15.1B, are

$$f_{cTi} = 0.55 \text{ MPa compression} \quad \text{O.K.} \quad \text{(1.6 MPa tension admissible)}$$

$$f_{cBi} = 20.51 \text{ MPa compression} \quad \text{O.K.} \quad \text{(24 MPa compression admissible)}$$

Concrete stresses with full superimposed load at the top of the beam are

$$f_{cTr} = \frac{P_r}{A} - \frac{P_r e}{S} + \frac{M_G}{S} + \frac{M_S}{S} = 18 \text{ MPa} \quad \text{(permissible compressive stress)}$$

$$= \frac{1.915}{0.225} - \frac{1.915 \times 0.250}{0.048\ 8} + \frac{0.105\ 5}{0.048\ 8} + \frac{M_S}{S} = 18$$

$$= 8.511 - 9.810 + 2.162 + \frac{M_S}{S} = 18$$

$$\frac{M_S}{S} = 17.137 \text{ MPa}$$

At the bottom of the beam,

$$f_{cBr} = 8.511 + 9.810 - 2.162 - \frac{M_S}{S} = -3.2 \quad \text{(permissible tensile stress)}$$

$$\frac{M_S}{S} = 19.359 \text{ MPa}$$

The stress at the top of the beam is critical:

$$M_S = 17.137 \times 0.048\ 8$$

$$= \tfrac{1}{8} \times w_S \times 10^{-3} \times 12.5^2$$

and the maximum permissible superimposed load is

$$w_S = 42.8 \text{ kN/m}$$

15.7 LOAD BALANCING

If the prestressing tendons are suitably shaped (Fig. 15.9) they can completely counteract the bending moment of a load system, and the beam is entirely free from bending moment. It is compressed along its axis, but this causes no deflection. In practice, the beam can be balanced for one load system only, and it is therefore subject to some, but a very much lower, bending moment under different load systems.

The beam carries a dead load due to its own weight and to other permanent loads. In addition there is a live load, which may or may not be acting. It is best to balance the full dead load and one-half of the live load (Fig. 15.10).

Load balancing is particularly useful when deflection is a problem. We noted in Sections 8.7 and 12.5 that creep deflection is a problem in many structural members, such as flat plates. The creep deflection is directly proportional to the elastic deflection; if the elastic deflection is zero, as it is in Figs. 15.9 and 15.10(b), the creep deflection also remains zero. In practice,

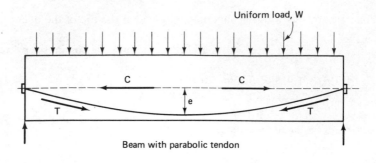

Beam with parabolic tendon

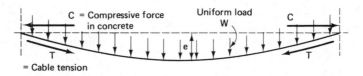

Parabolic suspension Cable

Figure 15.9 A beam carrying a uniformly distributed load over a simply supported span has a bending moment that varies parabolically along the span. We use a tendon of the same parabolic shape, with zero eccentricity at the supports, and a maximum eccentricity at midspan. If the moment due to the cable tension equals the bending moment $Pe = M = \frac{1}{8} W\ell^2$ at midspan, the bending moment due to the prestress is exactly equal and opposite to the bending moment due to the load along the entire span. The beam is subject only to a compressive force P. The action is similar to that in a suspension cable, which is subject to tension only.

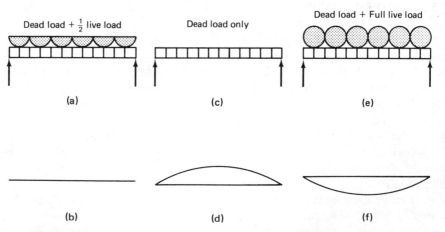

Figure 15.10 The live load may or may not act on the beam. It is therefore advisable to balance the load for the entire dead load and one-half of the live load (a, b). When there is no live load, the beam has a slight upward deflection (c, d); and when the full live is acting, it has a slight downward deflection (e, f).

the load cannot be perfectly balanced all the time, and there is some deflection, but it is smaller. Quite long spans are therefore possible with flat plates by prestressing them.

Example 15.3A

A rectangular prestressed concrete beam, 20 in. deep and 10 in. wide, spans 30 ft, and the tendons follow a parabolic profile, starting at middepth at each end, and reaching 7½ in. below the centerline at midspan. The beam is to be designed for zero bending moment under a load of 600 lb/ft. Determine the prestressing force required after loss of prestress. Calculate the stresses that would be caused by a load of 400 lb/ft, and by a load of 800 lb/ft.

    Data given: b = 10 in., h = 20 in., ℓ = 30 ft, e = 7½ in., w = 600, 400, or 800 lb/ft.

    Load balancing: The bending moment for the simply supported beam carrying a load of 600 lb/ft is

$$M = \tfrac{1}{8}\,w\ell^2 = 67,500 \text{ lb·ft} = 810,000 \text{ lb·in.}$$

The moment due to the prestressing force is

$$M = Pe = P \times 7.5 \text{ in.}$$

Therefore, the prestressing force required for load balancing is

$$P = \frac{810,000}{7.5} = 108,000 \text{ lb}$$

This requires 4 seven-wire strands.

    Different live loads: The cross-sectional area is

$$A = 10 \times 20 = 200 \text{ in.}^2$$

The section modulus is

$$S = \tfrac{1}{6} \times 10 \times 20^2 = 667 \text{ in.}^3$$

The midspan bending moment due to a load of 400 lb/ft is $M = \tfrac{1}{8} \times 400 \times 30^2 \times 12 = 540,000$ lb·in. and due to 800 lb/ft, $M = 1,080,000$ lb·in. The stresses due to a load of 400 lb/ft are

$$\frac{P}{A} \mp \frac{Pe}{S} \pm \frac{M}{S} = \frac{108,000}{200} \mp \frac{108,000 \times 7.5}{667} \pm \frac{540,000}{667}$$

$$= 540 \mp 1214 \pm 810$$

$$= 136 \text{ psi compression on top and 944 psi}$$
$$\text{compression on the bottom}$$

The stresses due to a load of 800 lb/ft are

$$540 \mp 1214 \pm 1619 = 945 \text{ psi compression top and 135 psi compression bottom}$$

This compares with a uniform compressive stress of 540 psi due to a load of 600 lb/ft.

Example 15.3B

A rectangular prestressed concrete beam, 500 mm deep and 250 mm wide, spans 10 m, and the tendons follow a parabolic profile, starting at middepth at each end, and reaching 200 mm below the centerline at midspan. The beam is to be designed for zero bending moment under a load of 8 kN/m. Determine the prestressing force required after loss of prestress. Calculate the stresses that would be caused by a load of 6 kN/m, and by a load of 10 kN/m.

Data given: $b = 250$ mm, $h = 500$ mm, $\ell = 10$ m, $e = 200$ mm, $w = 8$ kN/m, 6 kN/m, or 10 kN/m.

Load balancing: The bending moment for the simply supported beam carrying a load of 8 kN/m is

$$M = \tfrac{1}{8}\, w\ell^2 = 100 \text{ kN·m}$$

The moment due to the prestressing force is

$$M = Pe = P \times 0.20 \text{ m}$$

Therefore, the prestressing force required for load balancing is

$$P = \frac{100}{0.20} = 500 \text{ kN}$$

This requires 4 seven-wire strands.

Different live loads: The cross-sectional area is

$$A = 0.25 \times 0.5 = 0.125 \text{ m}^2$$

The section modulus is

$$S = \tfrac{1}{6} \times 0.25 \times 0.5^2 = 0.010 \; 42 \text{ m}^3$$

The midspan bending moment due to a load of 6 kN/m is $M = \tfrac{1}{8} \times 6 \times 10^2 = 75$ kN·m, and due to 10 kN/m, $M = 125$ kN·m. The stresses due to a load of 6 kN/m are

$$\frac{P}{A} \mp \frac{Pe}{S} \pm \frac{M}{S} = \frac{0.5}{0.125} \mp \frac{0.5 \times 0.2}{0.010 \; 42} \pm \frac{0.075}{0.010 \; 42}$$

$$= 4.00 \mp 9.60 \pm 7.20$$

$$= 1.6 \text{ MPa compression on top and 6.4 MPa}$$
$$\text{compression on the bottom}$$

The stresses due to a load of 10 kN/m are

$$4.00 \mp 9.60 \pm 12.00 = 6.40 \text{ MPa compression top and}$$
$$1.6 \text{ MPa compression bottom}$$

This compares with a uniform compressive stress of 4.0 MPa due to a load of 8 kN/m.

EXERCISES

1. What is meant by pretensioning and by posttensioning? Which types of structural member would be suitable for each of these prestressing techniques?
2. Why does prestressed concrete use less steel than normal reinforced concrete for the same loads and spans?
3. What are the causes of the loss of prestress (a) in posttensioned concrete and (b) in pretensioned concrete?
4. Why is the initial prestress the highest stress that will ever occur in the prestressing steel, irrespective of the load placed on a prestressed beam?
5. Why is the steel used for prestressing more commonly in the form of wires and strands than in the form of bars?
6. In reinforced concrete design we normally consider the ultimate strength of a member, but alternatively we could design the member by the elastic theory and consider its behavior under the action of the service loads. These are alternative methods, and either method gives a safe design. In prestressed concrete design, both the behavior under the action of the service loads *and* the ultimate strength must be checked. Why is that necessary?
7. Explain what is meant by the load balancing of prestressed concrete beams. How do we choose the load that is to be balanced?
8. Why can load balancing be used to counter creep deflection?

REFERENCES

15.1. H. J. Cowan, *Science and Building*, Wiley, New York, 1978.
15.2. T. Y. Lin, *Design of Prestressed Concrete Structures*, 2nd ed., Wiley, New York, 1963.
15.3. Committee on Prestress Losses, "Recommendations for Estimating Prestress Losses," *Journal of the Prestressed Concrete Institute*, Vol. 20, July–August 1975, p. 44.

Chapter 16

Concrete Shells and Folded Plates

The design of concrete shells and folded plates also deserves a textbook on its own. The mechanical advantages and general principles are explained, and some of the simpler structures are designed. These include two domes, a cylindrical shell, and a folded plate.

16.1 MECHANICAL ADVANTAGES AND PRACTICAL DISADVANTAGES OF CURVED AND FOLDED ROOFS

The principle of load balancing (Section 15.7) can also be applied to the shape of the roof structure (Fig. 16.1). Provided that the shape of a beam or slab equals that of the bending moment diagram, the structure is entirely in compression if it points upward. For concrete this is a particularly efficient solution, from a theoretical point of view. Thus a point load or a series of point loads require a folded-plate structure [Fig. 16.1(a)], a load uniformly distributed in plan requires a parabolic cylindrical barrel vault [Fig. 16.1(b)], and a curved roof of uniform thickness should be of catenary shape, that is, an upside-down suspension cable, to be free from bending moments due to its own weight [Fig. 16.1(c)].

Any variation in the distribution of the loads will introduce bending moments, but since the greatest load carried by a curved concrete roof is usually its own weight, these bending moments are unlikely to be large.

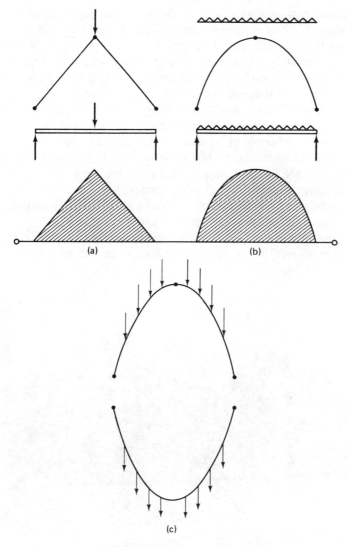

Figure 16.1 Moment-free arches.   (a) Linear arch carrying a concentrated load.
(b) Parabolic arch carrying a load uniformly distributed in plan.   (c) Catenary arch
carrying a load uniformly distributed along the length of the arch.   When extended
into the third dimension (a) becomes a folded-plate roof, and (b) and (c) become
barrel vault shell roofs.

In practice, curved or sloping *concrete* roof structures are not very common.   Sloping timber roofs are used on small buildings, because they shed
the rain, and tiles or shingles can be laid on a light structure.   A flat roof
requires more timber and more careful waterproofing.   Using concrete, the
sloping or curved roof structure costs more.

Concrete, however well made, is not a reliable material for keeping the rain out, although it is sometimes used without a waterproof covering, and such structures are sometimes waterproof.

The most reliable material for waterproofing is thin sheet metal, suitably jointed, but this is expensive, particularly for domes where much cutting is needed.   Various plastics are commonly used.

The formwork also adds to the cost.   Domes in particular present a problem, because their spherical surface cannot be produced from straight pieces of timber without a great deal of cutting.   Domes can, however, be formed on inflated membranes; the resulting dome is not spherical (Fig. 16.2).

Cylindrical shells and conoidal shells can be formed by straight pieces of timber resting on curved supporting frames.

The cost of forming folded plate structures has been greatly reduced by the development of waterproof plywood and hardboard.

The cost of the formwork is less significant if the same roof shape is repeated a number of times, so that the formwork can be reused.

Figure 16.2 Thin shell dome cast on a plastic balloon that is subsequently inflated while the concrete is still wet.   The reinforcement consists of steel springs placed diagonally that expand as the balloon is inflated.   After the concrete has hardened sufficiently, the balloon is deflated and removed for future use.   The necessary holes for doors and windows are then cut into the concrete.   An insulating layer is sprayed on both the outside and the inside of the concrete to give the dome the necessary insulation and stiffness.   Until the insulating layers have been placed on the dome, it is sensitive to temperature gradients.   The technique was devised by Dr. Dante Bini in Italy in the mid-1960s.   [Malvern Girls' College, Malvern, England, completed in 1978, with a diameter of 60 ft (18 m); photograph courtesy of Dr. Bini.]

16.2 MEMBRANE THEORY

The elements of the theory of shell roofs have been briefly discussed in several books (e.g., Ref. 7.2), and in more detail in several others (e.g., Ref. 16.2).

An element of a shell is subject to the forces and moments indicated in Fig. 16.3. In many thin reinforced concrete shells with a modest span the bending moment M, the twisting moment M_T, and the transverse shear forces Q can be neglected, except near the supports. The only forces acting on the shell are then the two forces N_x and N_y at right angles to one another, and the shear force V.

All these forces act within the surface of the shell, and they are called the membrane forces. A soap bubble, for example, can resist the three membrane forces. It is possible to push or pull it in two directions within the surface of the bubble at right angles to one another, provided that it is done carefully, and it is also possible to distort a soap bubble, that is, to shear it within its own surface.

The membrane forces can be worked out by statics alone. The solution may be found in any book on the theory of shells (e.g., Ref. 16.2, p. 18).

Reinforced concrete shells and folded plates are designed by the elastic theory. The stresses in the shell and in the edge members under the action of the service load are calculated. The compressive concrete stress is limited to $0.45f'_c$, and the tension is resisted by reinforcement. For Grade 40 steel the permissible steel stress is 20,000 psi and for Grade 60 steel it is 24,000 psi. It is preferable to use a Grade 40 steel since a stress of 20,000 psi produces a smaller extension of the concrete than one of 24,000 psi, so that it cracks less.

The minimum cover (*ACI Code*, Section 7.7.1) is ¾ in. for No. 6 bars or larger, ½ in. for No. 5 bars or smaller, and ½ in. for wire and welded mesh. This determines the minimum permissible thickness.

The maximum permissible spacing of the reinforcement (*ACI Code*, Section 19.4) is five times the thickness of the shell, or 18 in. Where the calculated tensile stress due to the factored load exceeds $4 \phi \sqrt{f'_c}$, the shell reinforcement must not be spaced farther apart than three times the shell thickness.

Figure 16.3 Internal forces and moments resisting the loads acting on a shell structure. *Forces acting within the surface of the shell (membrane forces):* N_x and N_y are direct (tensile or compressive) forces; V are the membrane shear forces. *Shear forces cutting through the shell: Q_x and Q_y are transverse shear forces. Moments: M_x and M_y are bending moments; M_{Tx} and M_{Ty} are twisting moments.*

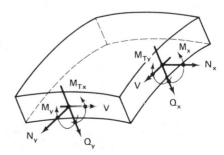

Shells and folded plates require shrinkage and temperature reinforcement (Section 6.3). A higher reinforcement ratio is required in the tensile zone, namely 0.003 5 times the gross cross-sectional area (*ACI Code*, Section 19.4.8).

16.3 DOMES

Domes are usually spherical, although that is not necessary. Some of the most famous domes, particularly those built during the Renaissance and Baroque, have been steeper, and this gives a more favorable stress distribution (Ref. 16.1, Chap. 7). However, the spherical dome has the simplest geometry, and this produces the simplest answer to the stress distribution within the shell.

It is convenient to consider vertical "great" circles, which correspond to the meridians of longitude on the earth's surface, and horizontal "small" circles, corresponding to the circles of latitude on the earth's surface.

It is also convenient to use spherical coordinates: the radius of curvature r, which is a constant for a spherical surface, a horizontal angle ϕ (at right angles to the section in Fig. 16.4), and a vertical angle θ, measured from the highest point of the dome (the crown). For a hemispherical dome $\theta = 90°$ at the supports of the dome (the springings).

In a spherical dome the shear forces V are zero, and there are therefore only two direct membrane forces, along the vertical meridian circles (N_θ) and along the horizontal latitude circles (N_ϕ). The meridional forces N_θ represent the arch action within the dome, which can be considered as formed by a series of interacting arches joined at the crown or north pole. The latitudinal or hoop forces change from compression near the crown to tension at the equator; the change occurs at angle (Fig. 16.4) $\theta = 51°50'$ from the crown. The two forces are (Ref. 16.2, p. 103):

$$N_\theta = -wr \frac{1}{1 + \cos \theta} \tag{16.1}$$

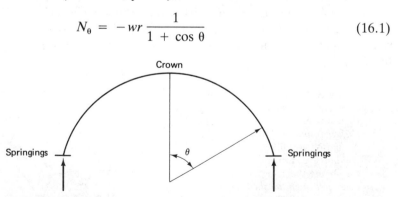

Figure 16.4   Coordinates for spherical dome: r = radius of curvature; θ = vertical angle with crown; ϕ = horizontal angle between meridians (not shown).

and

$$N_\phi = wr \left(\frac{1}{1 + \cos\theta} - \cos\theta \right) \qquad (16.2)$$

These forces have the values given in Table 16.1.

If the dome is hemispherical, its reaction comes vertically onto the springings, and the hoop forces restrain the arch action represented by the meridional forces.   The dome is therefore a stable structure without outward reactions.   In that respect it differs from the semicircular arch, which has horizontal reactions that must be restrained by abutments, or by a tie between the supports.

A hemispherical dome has an impressive height, particularly if it sits on top of a drum, but it contains a great deal of unusable space.   Even today, few people would seriously consider heating or air conditioning any of the great domed cathedrals of the Renaissance; but the interior of a modern dome generally requires at least heating, possibly air conditioning.   This is one of the main reasons why most modern domes are shallow.

The shallow dome was used in Byzantine and in some Muslim architecture.   Because the reactions are at an angle to the vertical, and because the tensile hoop forces are partly or wholly eliminated, there is an outward horizontal reaction, as there is in any arch.   This must be absorbed.   In Byzantine architecture buttresses, sometimes in the form of semidomes, were used, but a much simpler and more economical, if less dramatic, solution is the use of a tie ring (Fig. 16.5).   The tie has to absorb the horizontal component of arch action, that is, the forces

$$R_H = N_\theta \cos\theta \qquad (16.3)$$

which press outward in a circle from the edge of the dome.

The tie ring is placed in tension by the outward pressing forces R_H, and expands.   But if the dome is shallow, the hoop forces N_ϕ are compressive (see Table 16.1), and the edge of the dome contracts.   The tie ring is cast monolithically with the edge of the dome, and the edge of the dome contracts, while the tie expands.   This sets up bending stresses in the edge of the dome

TABLE 16.1   Values of Meridional Force N_θ
and Hoop Force N_ϕ in a Spherical Dome[a]

Angle with Crown (deg)	N_θ	N_ϕ
0	$-0.500wr$	$-0.500wr$
30	$-0.536wr$	$-0.330wr$
45	$-0.586wr$	$-0.121wr$
60	$-0.667wr$	$+0.167wr$
90	$-1.000wr$	$+1.000wr$

[a] $-$ denotes compression; $+$ denotes tension.

Figure 16.5 The horizontal component $R_H = N_\theta \cos \theta$ is in a shallow dome absorbed by a tie ring.

(Fig. 16.6). These bending stresses are confined to the lower part of the dome, and in a small dome it is not necessary to calculate them, provided that the edge of the dome is made much thicker than the shell, and suitably reinforced (Fig. 16.7).

Example 16.1A

Design a hemispherical dome, 3 in. thick with a span of 70 ft, using Grade 3.5 concrete, and Grade 40 steel.

Data given: $r = 35$ ft, $h = 3$ in., $f_c' = 3500$ psi, and $f_y = 40,000$ psi.
The concrete shells weigh

$$3 \times 12 = 36 \text{ psf}$$

A small additional dead load is required for a waterproof plastic coating. The live load must allow for the weight of a repair team. Wind pressure on a hemispherical surface is small; 18 psf is a conservative allowance for all these loads. We will therefore take $w = 54$ psf.

From Table 16.1 the maximum compressive force is

$$-1.0wr = 54 \times 35 = 1890 \text{ lb/ft} = 157.5 \text{ lb/in.}$$

and the maximum tensile force is

$$+1.0wr = 1890 \text{ lb/ft}$$

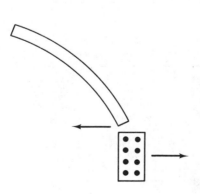

Figure 16.6 The hoop forces in a shallow dome are entirely compressive, so that the edge of the dome contracts. The meridional, or arch, forces are also compressive, and their horizontal component must be restrained by a tie ring which consequently expands. The expansion of the tie and the contraction of the edge of the shell are geometrically incompatible, and they produce bending stresses in the shell. The shell must therefore be made thicker near its edge, and given additional reinforcement to resist the bending moment.

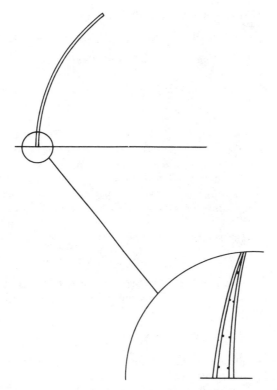

Figure 16.7 The design of the springings of a hemispherical shell depends on the restraint imposed by the supporting structure. The shell must be made thicker and given double reinforcement separated by a small moment arm to resist bending moments due to the support conditions.

The maximum compressive stress is

$$f_c = \frac{157.5}{1 \times 3} = 52.5 \text{ psi}$$

The maximum permissible stress is $0.45f'_c = 1575$ psi   O.K.

Tension reinforcement: The maximum permissible steel stress $f_s = 20{,}000$ psi. The area of reinforcement required for the hoop stress near the springings is

$$A_s = \frac{1890}{20{,}000} = 0.095 \text{ in.}^2/\text{ft}$$

The minimum area of reinforcement required to resist temperature and shrinkage stresses (Section 6.3) is

$$0.002 \times 12 \times 3 = 0.072 \text{ in.}^2/\text{ft}$$

The minimum area of reinforcement in the tension zone (Section 16.2) is

$$0.003\,5 \times 12 \times 3 = 0.126 \text{ in.}^2/\text{ft}$$

The maximum permissible bar spacing is the lesser of

$$5 \times 3 = 15 \text{ in.} \quad \text{or} \quad 18 \text{ in.}$$

Use No. 3 deformed bars Grade 40 for hoops at 10½-in. centers in the tension zone (0.13 in.²/ft).

The length of a quarter-circle of 35-ft radius is

$$½\pi \times 35 = 55.0 \text{ ft}$$

The tension zone extends over 90° − 51°50′ = 38°10′ along the meridians that is, over a distance of 55 × 38.17/90 = 23.4 ft. At 10½-in. centers, we require 27 hoops. *Use 27 No. 3 hoops at 10½ in. centers in the tension zone.*

Temperature and shrinkage reinforcement: For the remainder of the shell 0.072 in.²/ft reinforcement is required at a maximum spacing of 15 in. *Use No. 3 deformed bars at 15-in. centers for hoops above the tension zone and along the meridional lines* (0.09 in.²/ft).

Using No. 3 bars at right angles, the cover is

$$½ (3 - 2 \times 2 \times 0.375) = 0.75 \text{ in. (0.50 in. required)} \quad \text{O.K.}$$

At the springings of the dome the shell must be made thicker and given additional reinforcement. This depends on the restraint imposed by the supporting structure (Fig. 16.7).

Example 16.1B

Design a hemispherical dome, 75 mm thick, with a span of 20 m, using Grade 25 concrete, and Grade 300 steel.

Data given: $r = 10$ m, $h = 75$ mm, $f'_c = 25$ MPa, and $f_y = 300$ MPa.

The concrete shells weigh

$$0.075 \text{ m} \times 24 \text{ kN/m}^3 = 1.8 \text{ kPa}$$

A small additional dead load is required for a waterproof plastic coating. The live load must allow for the weight of a repair team. Wind pressure on a hemispherical surface is small. 1 kPa is a conservative allowance for all these loads. We will therefore take $w = 2.8$ kPa.

From Table 16.1 the maximum compressive force is

$$-1.0wr = 2.8 \times 10 = 28 \text{ kN/m}$$

and the maximum tensile force is

$$+1.0wr = 28 \text{ kN/m}$$

The maximum compressive stress is

$$f_c = \frac{28 \times 10^{-3}}{1.0 \times 0.075} = 0.37 \text{ MPa}$$

The maximum permissible stress is $0.45f'_c = 11.25$ MPa.   O.K.

Tension reinforcement: The maximum permissible steel stress $f_s = 140$ MPa.

The area of reinforcement required for the hoop stress near the springings is

$$A_s = \frac{28 \times 10^{-3}}{140} = 0.20 \times 10^{-3} \text{ m}^2 = 200 \text{ mm}^2/\text{m}$$

The minimum area of reinforcement required to resist temperature and shrinkage stresses (Section 6.3) is

$$0.002 \times 1\,000 \times 75 = 150 \text{ mm}^2/\text{m}$$

The minimum area of reinforcement in the tension zone (Section 16.2) is

$$0.003\,5 \times 1\,000 \times 75 = 262.5 \text{ mm}^2/\text{m}$$

The maximum permissible bar spacing is the lesser of

$$5 \times 75 = 375 \text{ mm} \quad \text{or} \quad 450 \text{ mm}$$

Use No. 10 deformed bars Grade 300 for hoops on 375-mm centers in the tension zone (267 mm²/m).

Example 16.2

Design the reinforcement for a shallow spherical dome, 3 in. thick with a span of 70 ft, using Grade 3.5 concrete and Grade 40 steel. The springings of the dome subtend an angle of 60° at the center of curvature.

Since the shell subtends an angle of 60° at the center of curvature (Fig. 16.8), the angle between the span ℓ and the radius of curvature r is also 60°, and the triangle ℓ, r, r is an equilateral triangle. Hence $\ell = r$.

Data given: $\ell = 70$ ft, $r = 70$ ft, $h = 3$ in., $f'_c = 3500$ psi, and $f_y = 40,000$ psi. Using the considerations set out in Example 16.1.A, we will take the load $w = 54$ psf.

Design of shell: The maximum compressive stresses occur at the springings, where $\theta = 30°$. From Table 16.1 the maximum compressive force is

$$-0.536wr = 2026 \text{ lb/ft} = 169 \text{ lb/in.}$$

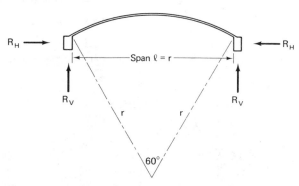

Figure 16.8 Forces at the springings of a shallow dome, subtending an angle of 60° at the center of curvature.

and the maximum compressive stress is

$$f_c = \frac{169}{1 \times 3} = 56 \text{ psi } (1575 \text{ psi is permissible}) \quad \text{O.K.}$$

There are no tensile stresses in the concrete. Following the considerations set out in Example 16.1, *Use No. 3 deformed bars at 15-in. centers for hoops and along meridians.*
    Design of tie: The meridional force at the springings, from Table 16.1, is

$$N_\theta = -0.536wr = 2026 \text{ lb/ft}$$

and its horizontal component is

$$R_H = 30 \cos 30° = 2026 \times 0.866 = 1755 \text{ lb/ft}$$

This presses outward on the circular tie, a problem mechanically analogous to that of a thin pipe, which is discussed in many textbooks of mechanics. Using the analogy of the pressure pipe, we cut the tie across a diameter (Fig. 16.9), and consider the equilibrium of the forces acting along that diameter. We have a uniform pressure of 1755 lb/ft resisted by the forces in each section of the tie cut by the diameter:

$$2T = 1755 \text{ lb/ft} \times 70 \text{ ft}$$

and the force in the tie is

$$T = 61,400 \text{ lb}$$

The maximum permissible steel stress $f_s = 20,000$ psi, and the area of reinforcement required is

$$A_s = \frac{61,400}{20,000} = 3.07 \text{ in.}^2$$

Use 4 No. 8 bars (= 3.16 in.²).
    The minimum space between No. 8 bars (Section 6.3) is $d_b = 1$ in. We require a No. 3 tie around the 4 bars, to prevent spalling of the cover as the No. 8 bars take up the tension. Thus the minimum width of the tie is

$$2 \times 1 \text{ (bar diameters)} + 1 \text{ (space)} + 2 \times 0.38 \text{ (tie)}$$

$$+ 2 \times 0.75 \text{ (cover)} = 5.25 \text{ in.}$$

Use a 6 in. by 6 in. tie.
    We gradually increase the shell thickness, so that it becomes 6 in. at the junction with the tie (Fig. 16.10).

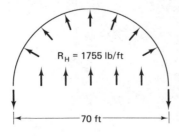

Figure 16.9 Forces acting on the tie of a shallow dome, using the pressure-pipe analogy.

$R_H = 1755 \text{ lb/ft}$

70 ft

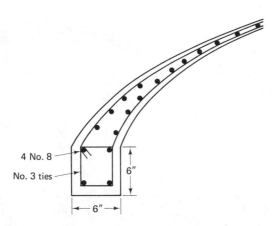

Figure 16.10   Reinforcement for the tie and the adjacent portion of the dome.

16.4 CYLINDRICAL SHELLS

Cylindrical shells are almost invariably built with a constant radius of curvature; that is, their surface forms part of a circular cylinder. It is usually less than a semicylinder both for reasons of appearance and of utility, since the semicircular cross section contains a large volume of air that has to be heated and cooled.

One cross section of the shell is curved, the other is straight. Its mechanical behavior is thus a blend of arch action due to the curved cross section, and of beam action due to the curved cross section spanning as a beam between the supporting columns.

If the width b (Fig. 16.11) is greater than twice the span ℓ, the arch action predominates, and the shell can be designed by the membrane theory (Section 16.2) except in region of the supports. If the span ℓ is greater than twice the width b, the beam action predominates. The membrane theory is then in error, but the shell can be designed as an ordinary reinforced concrete beam of curved cross section, provided that due allowance is made for the secondary stresses arising from its construction as a thin shell.

A cylindrical shell must be tied across its width, b, by a tie, a diaphragm, or a stiff frame. If that is not done, the shell flattens out and collapses. Stiffeners along the edges parallel to the span are useful for accommodating the large area of tension reinforcement needed if $\ell \gg b$, but they are not required for the equilibrium of the shell as are the ties parallel to b.

The fundamental equation of reinforced concrete design

$$M = f_s A_s z$$

applies also to shell construction. In shells where $\ell > b$, the height of the curved structure for a given span is less than in domes, and therefore the

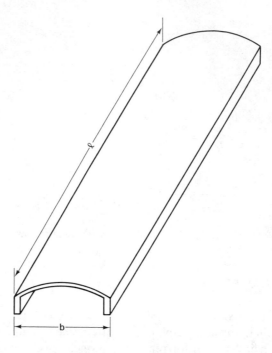

Figure 16.11 Span ℓ and width b of the cylindrical barrel vault shell. Ties (not indicated) are needed across the width b. Stiffeners parallel to the span (indicated) are useful for accommodating the tension reinforcement when $\ell \gg b$.

moment arm z does not benefit from the curvature as much as in a dome, although more so than in a beam-and-slab structure. Hence cylindrical shells need more reinforcement than do domes of the same span, but less than flat roofs of the same span.

The span of cylindrical shells is generally much greater than that of flat reinforced concrete roofs; that is, they have fewer supporting columns. The amount of reinforcement that has to be accommodated can therefore present practical problems. The area of reinforcement is greatly reduced if prestressing tendons are used near the edges of a cylindrical shell. This is partly due to the much higher steel stress permissible, and partly due to the fact that the prestressing can be used to balance the load due to the weight of the shell (Section 15.7). If the tendons are curved, the shell is lifted off its formwork as it is prestressed, and it thus begins to carry its own weight. The prestress can be designed so that the weight of the shell, but not the live load, is balanced by the prestress, thus reducing both concrete stresses and long-term deflection.

A shear force equal to approximately half the weight of the shell occurs in the portion of the shell near the columns. This produces diagonal tension, which must be resisted by shear reinforcement. Stirrups cannot be used in shells, but bars from the edge stiffeners can be bent up at 45° into the shell. The large amount of reinforcement needed to resist the bending moment at midspan is available to bend up into the shell at 45°, and act as shear reinforcement.

Example 16.3A

Determine approximately the amount of tension reinforcement needed at midspan for the simply supported shell shown in Fig. 16.12A, using Grade 3.5 concrete and Grade 40 steel.

Data given: Span ℓ = 150 ft, width b = 50 ft, overall height h = 15 ft, thickness of shell h_f = 3 in., f'_c = 3500 psi, and f_y = 40,000 psi.

The complete design of this shell is beyond the scope of this elementary book. It either requires an exact analysis using the bending theory of cylindrical shells (Ref. 16.3), or an approximate analysis which assumes that the shell is a simply supported beam of the cross section shown. This gives a good approximation, if the centroid and moment of inertia are determined accurately. Computer programs are available for both methods.

In this simple example we will assume the length of the moment arm as two-thirds of the overall depth of the shell. This is too great an approximation for a proper design, but it serves to indicate the size of the edge beams needed.

The weight of a 3-in. thick shell is

$$3 \times 12 = 36 \text{ psf}$$

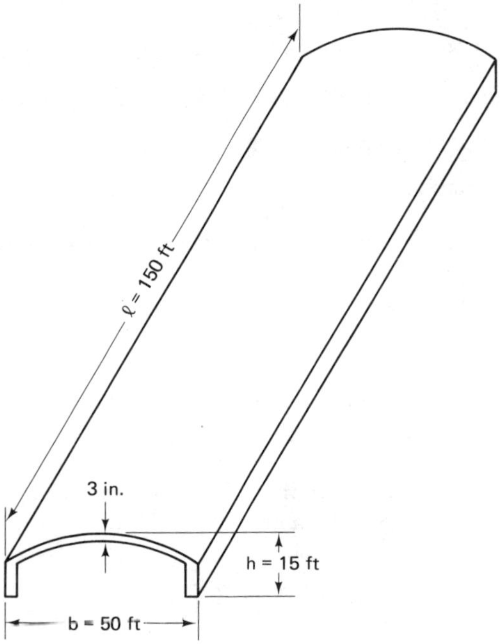

Figure 16.12A   Dimensions of the shell for Examples 16.3A and 16.4.

The curved shell weighs a little more if the weight is measured on plan. We must also allow for the weight of the edge stiffeners and the waterproofing membrane. This suggests a dead load of 44 psf. The live load must allow for the weight of a repair team. Wind pressure on cylindrical shells is small. A combined load of 60 psf is conservative.

The total load

$$W = 60 \times 50 \times 150 = 450,000 \text{ lb}$$

and the maximum bending moment for simple supports is

$$M = \tfrac{1}{8} \times 450,000 \times 150 = 8,438,000 \text{ lb} \cdot \text{ft}$$

The lever arm is assumed to be $z = \tfrac{2}{3} h = \tfrac{2}{3} \times 15 = 10$ ft, and the maximum tensile force is

$$T = \frac{M}{z} = \frac{8,438,000}{10} = 844,000 \text{ lb}$$

divided between two edge beams, or 422,000 lb per edge beam.

For a maximum permissible steel stress of 20,000 psi,

$$A_s = \frac{422,000}{20,000} = 21.1 \text{ in.}^2$$

This is a very large steel area, but the use of Grade 60 steel would produce more cracking. On the other hand, large-diameter bars can be used provided that their anchorage is carefully checked. Because of the long span, this should present no problem. *Use 15 No. 11 bars* ($= 23.4 \text{ in.}^2$).

Using an array 3 bars wide by 5 bars deep, the minimum width is

$$2 \times 0.75 \text{ (cover)} + 2 \times 0.38 \text{ (tie)} + 3 \times 1.41 \text{ (bars)} + 2 \times 1.41 \text{ (spaces)}$$

$$= 9.3 \text{ in.}$$

and the minimum depth is

$$2 \times 0.75 + 2 \times 0.38 + 5 \times 1.41 + 4 \times 1.41 = 14.94 \text{ in.}$$

Use 10 in. by 15 in. stiffeners.

Example 16.3B

Determine approximately the amount of tension reinforcement needed at midspan for the simply supported shell shown in Fig. 16.12B, using Grade 25 concrete and Grade 300 steel.

Data given: Span $\ell = 45$ m, width $b = 15$ m, overall height $h = 4.5$ m, thickness of shell $h_f = 75$ mm, $f'_c = 25$ MPa, and $f_y = 300$ MPa.

The weight of a 75-mm-thick shell is

$$0.075 \times 24 = 1.8 \text{ kPa}$$

The curved shell weighs a little more if the weight is measured on plan. We must also allow for the weight of the edge stiffeners and the waterproofing membrane. This suggests a dead load of 2.2 kPa. The live load must allow for the weight of a

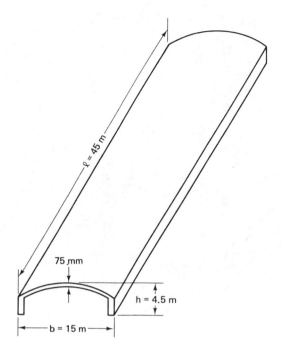

Figure 16.12B   Dimensions of the shell for Example 16.3B.

repair team.   Wind pressure on cylindrical shells is small.   A combined load of 3 kPa is conservative.

The total load

$$W = 3 \times 15 \times 45 = 2\ 025 \text{ kN}$$

and the maximum bending moment for simple supports

$$M = \tfrac{1}{8} \times 2.025 \times 45 = 11.4 \text{ MN} \cdot \text{m}$$

The lever arm is assumed to be $z = \tfrac{2}{3}h = \tfrac{2}{3} \times 4.5 = 3$ m, and the maximum tensile force is

$$T = \frac{M}{z} = \frac{11.4}{3} = 3.8 \text{ MN}$$

divided between two edge beams, or 1.9 MN per edge beam.   For a maximum permissible steel stress of 140 MPa,

$$A_s = \frac{1.9}{140} = 13.571 \times 10^{-3} \text{ m}^2 = 13\ 571 \text{ mm}^2$$

Use 15 No. 35 bars (= 15 000 mm²).
   Using an array 3 bars wide by 5 bars deep, the minimum width is

$$2 \times 20 \text{ (cover)} + 2 \times 11.3 \text{ (tie)} + 3 \times 35.7 \text{ (bars)} + 2 \times 35.7 \text{ (spaces)}$$

$$= 241 \text{ mm}$$

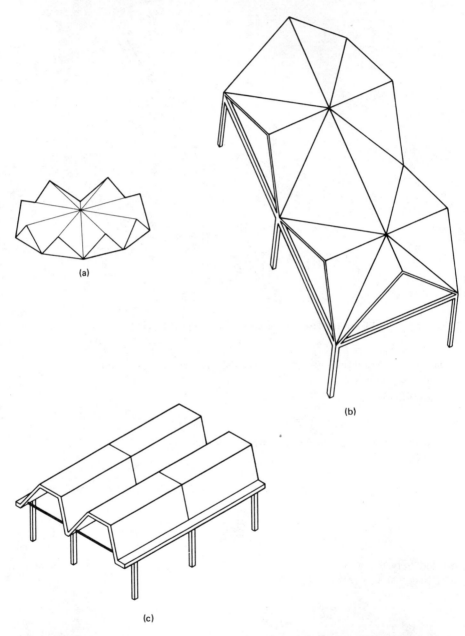

(a)

(b)

(c)

Figure 16.13 Some folded plate structures perform functions similar to those of shell structures. The following may be used instead: (a) a dome on a circular plan; (b) a dome on a square plan; (c) a cylindrical shell; and (d) a cylindrical shell with a north light for indirect lighting. Part (e) shows a folded-plate structure that has no equivalent in shell construction.

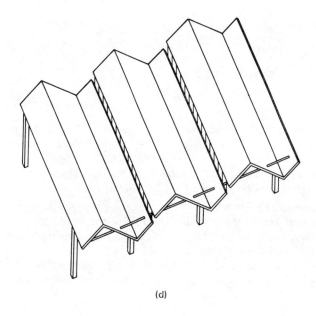

(d)

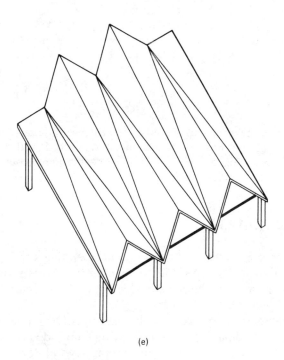

(e)

Figure 16.13   Continued.

and the minimum depth is

$$2 \times 20 + 2 \times 11.3 + 5 \times 35.7 + 4 \times 35.7 = 384 \text{ mm}$$

Use 250-mm by 400-mm stiffeners.

Example 16.4

Determine approximately the amount of prestressed steel needed at midspan for the simply supported shell shown in Fig. 16.12.A, using Grade 3.5 concrete and prestressing tendons with an ultimate steel strength of 270,000 psi.

Data given: Span $\ell = 150$ ft, width $b = 50$ ft, overall height $h = 15$ ft, thickness of shell $= 3$ in., $f'_c = 3500$ psi, and $f_u = 270,000$ psi.

The maximum eccentricity of the prestressing tendons depends on the size of the edge stiffeners and of the resulting centroid of the cross section. We have determined neither, but an eccentricity of 5 ft is obtainable for the shell shown in Fig. 16.12.A, even if the edge stiffeners are smaller.

We will balance the dead load of 44 psf determined in Example 16.3. Hence the maximum bending moment to be balanced by the parabolic cables is

$$M = \tfrac{1}{8} \times 44 \times 50 \times 150^2 = 6,187,500 \text{ lb} \cdot \text{ft}$$

$$= Pe = P \times 5 \text{ ft}$$

The maximum permissible steel stress is $0.70 f_u = 189,000$ psi (*ACI Code*, Section 18.5). We will allow 36,000 psi for loss of prestress (Section 15.3). Therefore, $f_s = 189,000 - 36,000 = 153,000$ psi. Consequently, we require an area of prestressing steel of

$$A_s = \frac{6,187,000}{5 \times 153,000} = 8.08 \text{ in.}^2$$

This area is divided between the two edges, that is, we require 4.04 in.^{2} each edge, or 20% of the steel area needed for a normally reinforced shell; however, the calculation is only approximate. The prestressing steel required could be provided by 19 seven-wire strands per edge beam.

16.5 FOLDED PLATES

Any sharp corner in a structure must produce bending stresses. Thus folded plates, unlike thin shells, are subject to bending. Their formwork, however, is much cheaper, because the flat surfaces can be constructed from hardboard or waterproof plywood.

The great increase in strength obtained by folding a flat surface is easily demonstrated by folding a sheet of paper which unfolded is unable to support its own weight. After folding it can support several dozen sheets of paper if the folds are prevented from straightening out. Thus a folded plate, like a cylindrical shell, needs ties (Fig. 16.13).

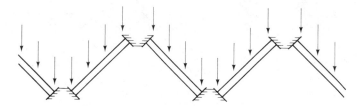

Figure 16.14 Solution of folded plate by moment distribution. The structure is dissected at the joints, which are firmly clamped. The joints are then released in turn, and the bending moment distributed until the residual differences are negligibly small.

Since folded plates are subject to bending moments, they need to be thicker than membrane shells *for the same span*. In practice, folded plates are particularly useful for roofs of buildings of more modest spans.

A folded plate is a rigid frame in three dimensions. It can be analyzed by any of the standard methods for designing rigid frames, such as moment distribution (Fig. 16.14) or the matrix displacement method (Section 7.3).

Many folded plate structures can, however, be designed by the method we employed in Example 16.3. If the structure has a linear symmetry, as in Fig. 16.13(c), (d) and (e), it can be considered as a beam, simply supported or continuous, and designed accordingly. Thus Fig. 16.15 shows a cross section through the folded plate shown in Fig. 16.13(e) at one of its supports; this explains its behavior as a normal reinforced concrete slab of unusual cross section, but a reinforced concrete slab nevertheless. It is necessary to determine the location of the neutral axis, which presents some problems, and is better done for the service-load condition.

Example 16.5

Determine the reinforcement required for the 4-in.-thick folded-plate cantilevered awning shown in Fig. 16.16. Use Grade 3.5 concrete and Grade 40 steel.

Data given: $\ell = 10$ ft, overall depth $h = 30$ in., slab thickness $h_f = 4$ in., $f'_c = 3500$ psi, and $f_y = 40,000$ psi.

This is a small structure, and approximations can safely be made without significant loss of economy, provided that the assumptions are conservative. The con-

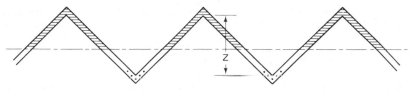

Figure 16.15 A cross section through the folded plate roof shown in Fig. 16.13(e) at its support.

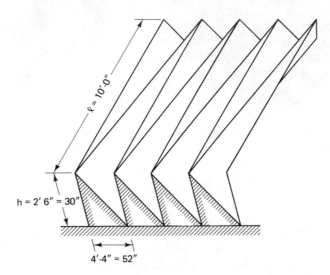

Figure 16.16 Dimensions of the cantilevered awning designed in Example 16.5.

crete plate weighs $4 \times 12 = 48$ psi per unit surface area, but it is built at an angle to the horizontal of $\tan^{-1}(30/26) = 49°$. Thus we must multiply by $\sec 49° = 1.5$.

We must allow for a plastic waterproof finish, for the live load caused by a possible repair team, and for a snow load. Assuming that the structure is to be built in a region of moderate snow fall, we will allow 32 psf. Wind load may cause an uplift, but not a downward pressure. This gives

$$w = 1.5 \times 48 + 32 = 104 \text{ psf}$$

The total load is

$$W = 104 \times 4.33 \times 10 = 4503 \text{ lb}$$

and the maximum bending moment is

$$M = -\tfrac{1}{2} W\ell = \tfrac{1}{2} \times 45.3 \times 10 = 22{,}516 \text{ lb} \cdot \text{ft}$$

At the service load the neutral axis of the reinforced concrete section is approximately at half the effective depth. Therefore, taking the moment arm z as half the overall depth is conservative.

$$z = \tfrac{1}{2} h \text{ approximately} = 15 \text{ in.} = 1.25 \text{ ft}$$

The maximum permissible steel stress $f_s = 20{,}000$ psi. The area of reinforcement required is

$$A_s = \frac{M}{f_s z} = \frac{22{,}516}{20{,}000 \times 1.25} = 0.90 \text{ in.}^2$$

Use 5 No. 4 bars (1.00 in.²), which are required for each triangle of the awning, at the top for a negative bending moment.

EXERCISES

1. Why is it possible to effect a great reduction in the bending moments, and thus use far less structural material, by using shells or folded plates in place of a combination of vertical and horizontal members?

2. Why are reinforced concrete shell structures and folded plates not very common, in spite of the evident mechanical advantages?

3. What are the assumptions made in membrane theory?   Which types of shell structure satisfy them sufficiently closely for use in design?

4. What is the function of the circular tie commonly used at the base of a shallow spherical dome?   Why does it set up bending stresses in the adjacent part of the shell?

5. Why is it necessary to tie a cylindrical shell across its width by a tie, a diaphragm, or a stiff frame?

6. Explain how a cylindrical shell whose length is much greater than its width can be designed as a "beam" of curved cross section.   Why is prestressing an advantage for these shells?

7. Describe the principal types of folded-plate structures.   Which of these can be designed as "beams"?

REFERENCES

16.1. Jürgen Joedicke, *Schalenbau*, Krämer, Stuttgart, 1962.
16.2. Alf Pflüger, *Elementary Statics of Shells*, Dodge, New York, 1961.
16.3. A. M. Haas, *Thin Concrete Shells*, 2 vols., Wiley, New York, 1962, 1967.

Appendix A

Glossary

Words in italics denote a cross reference.

Balanced Failure Condition. The simultaneous occurrence of a *primary tension failure* and a *primary compression failure*.

Cast-in-Place Concrete. Concrete cast on the building site in its final position, as opposed to *precast concrete*. See also *ready-mixed concrete*.

Column Strip. A strip of slab on either side of the column. It is one-half of the width of a flat slab or plate. See also *middle strip*.

Compression Failure, Primary. See *primary compression failure*.

Compression Reinforcement. Reinforcement stressed in compression, not in tension.

Cover. Concrete placed over the reinforcement to protect it from corrosion and fire.

Creep. Long-term deformation caused by squeezing of water from the fine pores of the concrete by sustained loads.

Crown. The highest point of an arch or dome.

Dead Load. Permanently applied load due to the weight of the structure, the cladding, the finishes, permanent partitions, and so on, as opposed to the *live load*.

Development Length.   The length of bar required to develop the stress at the critical section.

Effective Depth.   Denoted by d.   Distance of center of tension reinforcement from the compression face.   When one layer of reinforcement is used, it is the overall depth minus the bar radius minus the *cover*.

Elastic Analysis.   An analysis based on the elastic theory, as opposed to limit design, which is based on the plastic theory.

Elastic Deflection.   Deflection that occurs instantaneously as the load is applied, and is recovered when the load is removed, as opposed to *creep* deflection.

Elasticity, Modulus of.   See *modulus of elasticity*.

Equivalent Rectangular Stress Block.   A rectangle that has the same area and the same centroid as the experimentally determined *stress block*.

Factored Load.   See *fully factored load* and *partially factored load*.

Fixed-Ended.   Rigidly restrained at the supports.

Flat Plate.   A slab directly supported on the columns without column heads, as opposed to a *flat slab* and a *two-way slab*.

Flat Slab.   A slab directly supported on the columns, but joined to them through enlarged column heads to improve the shear resistance.

Formwork.   The mold in which the liquid concrete is placed; it gives its form to the concrete.

Fully Factored Load.   Load that has been multiplied by the load factor and divided by the strength-reduction factor.

Grade 3 (3.5, 4, etc., to 10) Concrete.   Concrete with a specified compressive strength of 3000 (3500, 4000, etc., to 10,000) psi.

Grade 20 (25, 30, etc., to 70) Concrete.   Concrete with a specified compressive strength of 20 (25, 30, etc., to 70) MPa.

Grade 40 (60) Steel.   Steel with a minimum yield stress of 40,000 (60,000) psi.

Grade 300 (400) Steel.   Steel with a minimum yield stress of 300 (400) MPa.

Hook.   A standard anchorage consisting of a bend followed by a straight length.

Hoop Forces.   The horizontal membrane forces in a dome.

Inflection, Point of.   A point where the bending moment is zero and the curvature changes from convex to concave.

L-Beam. A *T-beam* at the edge of a structural slab.

Lightweight Aggregate. An aggregate lighter than *normal aggregate*, used to reduce the weight of the concrete.

Lightweight Concrete. Concrete lighter than concrete made with *normal aggregate*. It may be made with *lightweight aggregate*, or with air or gas bubbles without any aggregate.

Live Load. Nonpermanent load due to the contents of the building and the people in it, as opposed to *dead load*.

Load Balancing. *Prestressing* a beam or slab so that it is subject to zero bending moment under its *service load*.

Load Factor. Factor by which each load is multiplied in ultimate strength design. See also *strength-reduction factor*.

Long-Term Deflection. Total deflection over a long period of time. It consists of both the *elastic deflection* and the *creep deflection*.

Membrane Forces. Direct and shear forces which can act entirely within a thin shell.

Meridional Forces. The vertical membrane forces in a dome, which act along lines corresponding to the meridians of longitude on the earth's surface.

Microcrack. A crack too fine to be visible with the naked eye but detectable with ultrasonic pulses.

Middle Strip. The center half of a flat slab or plate. See also *column strip*.

Modulus of Elasticity. Denoted by E. In elastic materials, the ratio of *stress* to *strain*. The higher the modulus of elasticity, the less the elastic deformation.

Moisture Movement. Deformation, such as shrinkage or swelling, caused by the absorption or evaporation of water or water vapor.

Moment Arm. Denoted by z. The distance between the resultant tensile force T and the resultant compressive force C.

Moment Arm Ratio. The ratio of the *moment arm* to the *effective depth*.

Moment Coefficient. A number, sometimes empirical, which gives the bending moment when multiplied by $w\ell^2$, where w is the uniformly distributed load and ℓ is the span.

Moment of Inertia. Denoted by I. A geometric property of the cross section, required for calculations of deflection and stiffness. It is defined as $\int by^2 \, dy$, and its value for a plain, rectangular concrete section is $\frac{1}{12} bh^3$.

Monolithic. Literally, one piece of stone. Denotes concrete cast all in one piece, or concrete from several casting operations built with proper construction joints.

Negative Bending Moment. Bending moment that causes tension on the top face and compression on the bottom face. It occurs in cantilevers and over the supports of continuous or restrained beams or slabs.

Negative Tension Reinforcement. Tension reinforcement on the top face of a beam or slab, which resists the *negative bending moment*.

Neutral Axis. Axis in a beam, slab, or eccentrically loaded column where the strain and stress changes from compressive to tensile. At the neutral axis the strain and stress are zero.

Normal Aggregate. Concrete aggregate consisting of crushed rock or gravel, as opposed to *lightweight aggregate*.

No. 3 (4, 5, 6, 7, 8, 9, 10, 11, 14, 18) Bar. A reinforcing bar with a nominal diameter of 3 (4, 5, etc.) 8ths of an inch; that is, a No. 4 bar is a ½-inch bar, a No. 8 is a 1-inch bar.

No. 10 (15, 20, 25, 30, 36, 45, 55) Bar. A metric reinforcing bar with a nominal diameter of 10 (15, 20, etc.) millimeters.

One-Way Slab. Slab designed to span in one direction only, as opposed to a *two-way slab*.

Partially Factored Load. Load that has been multiplied by the load factor but not divided by the strength-reduction factor.

Pile Cap. A reinforced concrete structure that distributes the load from a column to the piles that support it.

Plastic Deformation. The permanent, nonrecoverable deformation of steel above the *yield stress* or *proof stress*, as opposed to elastic deformation.

Portland Cement. The most common type of cement, made from limestone or chalk and shale or clay.

Positive Bending Moment. Bending moment that causes compression on the top face and tension on the bottom face. It occurs in simply supported beams and slabs, and at midspan in continuous or restrained beams or slabs.

Positive Tension Reinforcement. Tension reinforcement on the bottom face of a beam or slab, which resists the *positive bending moment*.

Posttensioned Concrete. *Prestressed concrete* whose steel is prestressed after the concrete has hardened.

Precast Concrete. Concrete cast in a position other than its final position.

It could be cast on the building site, but usually precasting is done in a factory.

Prestressed Concrete.   Concrete that is precompressed in the zone where tensile stresses occur under load; consequently, cracking of the concrete due to tension is avoided.

Pretensioned Concrete.   *Prestressed concrete* whose steel is prestressed before the concrete is cast.

Primary Compression Failure.   Failure of reinforced concrete initiated by crushing of the concrete.

Primary Tension Failure.   Failure of reinforced concrete initiated by *plastic deformation* of the steel.   Crushing of the concrete occurs subsequently and produces the actual failure.

Ready-Mixed Concrete.   Concrete mixed in a factory and then transported to the site.

Rectangular Stress Block, Equivalent.   See *equivalent rectangular stress block*.

Reinforcement Ratio.   Denoted by ρ.   The ratio of the area of reinforcement to the gross concrete area (in columns) or to the "effective" area bd (in beams and slabs), where b is the width and d is the *effective depth*.

Rigid Frame.   A frame with rigid joints.   In the elastic analysis of a rigid frame it is assumed that right-angled rigid joints remain right-angled when the frame is loaded.   The deformation is assumed to occur entirely in the members connecting the joints.

Secondary Reinforcement.   Reinforcement lighter than the main reinforcement, and generally at right angles to it.   It is required in slabs to distribute the load and to counter stresses due to temperature and shrinkage.   It is required in beams and columns to resist shear stresses and to counter buckling of the longitudinal reinforcement.

Section Modulus.   Denoted by S.   The moment of inertia divided by the distance of the top or bottom face from the *neutral axis*.

Service Load.   The actual load assumed to be acting on the structure.   Also called working load.

Short Column.   A column whose *slenderness ratio* is so low that its effect can be ignored.

Short-Term Deflection.   Primarily elastic deflection which occurs after a short period of time, as opposed to *long-term deflection*.

Shrinkage.   The contraction of concrete caused by evaporation and by the chemical reaction between cement and water.   In normal concrete the contraction amounts to approximately 3×10^{-4}, or one part in 3000.

Shrinkage Reinforcement. *Secondary reinforcement* designed to resist the tensile stresses caused when *shrinkage* contraction is restrained.

Slenderness Ratio. The ratio of the effective length to the radius of gyration or width, which determines the tendency of columns to buckle. A column with a high slenderness ratio is called a long column, and a column with a slenderness ratio so low as not to affect its strength is called a *short column*.

Specified Strength of Concrete. Denoted by f'_c. The compressive strength specified by the designer. The average strength of the concrete must be substantially higher, to an extent depending on the quality control (see Section 5.8).

Spirally Reinforced Column. A column wound with a continuous spiral, to prevent buckling of the longitudinal bars and expansion of the concrete core, as opposed to a *tied column*.

Strain. Denoted by ε. Deformation per unit length.

Strand. *Tendon* produced by twisting together several (usually seven) high-tensile wires.

Strength-Reduction Factor. Denoted by ϕ. A coefficient providing for the possibility that small adverse variations in the strength of the materials, in workmanship, and in dimensions, although individually within acceptable limits, may combine to lower the strength of the section below the acceptable level. See also *load factor*.

Stress. Denoted by f. Force per unit area.

Stress Block. The shape of the *stress–strain diagram* of concrete. It is proportional to the resultant compressive force when the reinforced concrete beam fails. It is usually idealized to an *equivalent* rectangle.

Stress–Strain Diagram. The experimental relation between stress and strain (for either steel or concrete) up to failure. Usually, the strain is plotted as the horizontal coordinate, and the stress as the vertical coordinate.

T-beam. A beam incorporating a portion of the slab that it supports.

Temperature Movement. Deformation caused by changes in temperature.

Temperature Reinforcement. Reinforcement designed to resist tensile stresses caused by *temperature movement*. It is usually combined with *shrinkage reinforcement*.

Tendon. A generic term for high-tensile bars, wires, strands, or cables used for *prestressing* concrete.

Tension Failure, Primary. See *primary tension failure*.

Tied Column. A column that uses ties to prevent buckling of the longitudinal reinforcement, as opposed to a *spirally reinforced column*.

Twisting Moment. Torsional moment.

Two-Way Slab. A slab supported on beams whose reinforcement is designed to span in both directions. The ratio of the longer to the shorter side must be less than 2 for two-way action to occur. A two-way slab supported directly on the columns is called a *flat plate*.

Ultimate Load. The actual working load multiplied by the *load factor*. The ultimate load does not include the *strength-reduction factor*.

Ultimate Strain. The strain at which a material fails. The ultimate strain of concrete is assumed to be the same for all types of concrete, and equal to 3×10^{-3}.

Ultimate Stress. The stress at which a material fails.

Working Load. See *service load*.

Yield Strain. The strain at which steel begins to show substantial plastic deformation.

Yield Stress. Denoted by f_y. The stress corresponding to the *yield strain*.

Appendix B

Notation

a	depth of equivalent stress block of concrete
A	area; cross-sectional area
A_g	gross area of cross section
A_s	area of tension reinforcement
A'_s	area of compression reinforcement
A_{sB}	area of bottom (positive) reinforcement
A_{st}	area of vertical column steel
A_{sT}	area of top (negative) reinforcement
A_v	area of shear reinforcement within a distance s
b	width of compression face of member
b_0	perimeter of critical section for shear in slabs and footings
b_w	width of web
B	width of footing
c	distance from extreme compression fiber to neutral axis
c_b	distance of extreme compression fiber to neutral axis which produces balanced strain conditions
c_{max}	$= 0.75c_b$
C	compressive force
d	effective depth

d'	distance of extreme compression fiber to centroid of compression reinforcement
d_b	nominal diameter of reinforcing bar
D	service (unfactored) dead load
e	eccentricity of load parallel to axis of member, measured from centroid of gross cross section
E	modulus of elasticity
E_c	modulus of elasticity of concrete
E_s	modulus of elasticity of steel
f	direct stress
f_c	actual compressive stress in concrete due to service (unfactored) load
f'_c	specified compressive strength of concrete
f_{ct}	tensile strength of concrete
f_h	tensile stress developed by a standard hook
f_s	actual stress in tension steel
f_t	tensile stress in concrete
f_u	ultimate strength of reinforcing steel
f_v	diagonal tensile stress due to shear
f_y	specified yield strength of reinforcement
g	acceleration due to gravity
h	overall depth; height
h_f	slab (flange) thickness
H	horizontal force
I	moment of inertia
k	deflection constant
k_m	design constant for rectangular beams and slabs for maximum permissible (ultimate load) design
ℓ	span
ℓ_a	additional embedment length at support or at point of inflection
ℓ_d	development length
ℓ_l	longer span of two-way slab
ℓ_n	clear span
ℓ_s	shorter span of two-way slab; clear distance between webs
L	service (unfactored) live load
m	moment factor (Section 7.3)
M	bending moment
M_-	negative bending moment

M_+	positive bending moment
M'	ultimate resistance moment of reinforced concrete section
M'_{max}	ultimate resistance moment of reinforced concrete section with maximum permissible reinforcement ratio
M_G	moment due to (service) weight of prestressed concrete beam
M_S	moment due to superimposed service load
M_x, M_y	bending moments along the x- and y-axes
M_u	ultimate (fully factored) bending moment
N_x, N_y	direct membrane forces
N_θ	meridional force in dome
N_ϕ	hoop force in dome
P	service (unfactored) column load
P_i	initial prestress
P_o	maximum concentric column load
P_r	prestress after losses
P_u	ultimate (fully factored) column load
q_s	average permissible soil pressure
Q_x, Q_y	transverse shear forces in shells
r	radius
R	reaction
R_H	horizontal reaction
R_V	vertical reaction
s	stirrup spacing; shrinkage
S	section modulus
T	tensile force
T_n	nominal (partially factored) torsional moment strength
T_u	ultimate (fully factored) torsional moment at section
U	ultimate (fully factored) load
v	unit shear stress
v_n	$= V_n/bd$
v_s	contribution of shear reinforcement to unit shear capacity
V	shear force
V_c	nominal shear strength provided by concrete
V_n	nominal (partially factored) shear strength
V_s	nominal shear strength provided by steel
V_t	shear force equivalent of torsional moment (Section 11.7)
V_u	ultimate (fully factored) shear force
w	service load per unit length, or per unit area

w_S	service superimposed load per unit area
w_u	ultimate load per unit length, or per unit area
W	total service load; service wind load
x	distance parallel to span; general variable
$\bar{x}$	average value of x
y	vertical distance
z	arm of resistance moment
α	ratio of the concrete strength in a beam to its cylinder crushing strength f'_c
α_l, α_s	bending moment coefficients for two-way slabs
β_1	factor used in the equivalent rectangular stress diagram for concrete at the ultimate load. It is 0.85 for f'_c up to 4000 psi; for concrete strength above 4000 psi it is reduced at the rate of 0.08 for each 1000 psi of strength in excess of 4000 psi
γ_c	density of concrete
γ_f	unit weight of soil
Δ	elastically computed deflection
ε	strain
ε_c	compressive strain in concrete
ε'_c	ultimate concrete strain $= 0.003$
ε_s	strain in tension steel
ε'_s	strain in compression steel
ε_y	yield strain of steel $= f_y/E_s$
θ	vertical angle
π	circular constant $= 3.142$
ρ	ratio of tension reinforcement to effective concrete section
ρ'	ratio of compression reinforcement to effective concrete section
ρ_b	reinforcement ratio producing balanced strain conditions
ρ_{max}	$= 0.75\rho_b$
ρ_t	$= A_{st}/A_g$
σ	standard deviation
ϕ	strength-reduction factor; horizontal angle

Appendix C

Tables of Reinforcing Bar Areas

TABLE C.1 Cross-Sectional Area (in.²) of Combinations of Bars of the Same Size

Number of Bars	Bar Number										
	3	4	5	6	7	8	9	10	11	14	18
1	0.11	0.20	0.31	0.44	0.60	0.79	1.00	1.27	1.56	2.25	4.00
2	0.22	0.40	0.62	0.88	1.20	1.58	2.00	2.54	3.12	4.50	8.00
3	0.33	0.60	0.93	1.32	1.80	2.37	3.00	3.81	4.68	6.75	12.00
4	0.44	0.80	1.24	1.76	2.40	3.16	4.00	5.08	6.24	9.00	16.00
5	0.55	1.00	1.55	2.20	3.00	3.95	5.00	6.35	7.80	11.25	20.00
6	0.66	1.20	1.86	2.64	3.60	4.74	6.00	7.62	9.36	13.50	24.00
7	0.77	1.40	2.17	3.08	4.20	5.53	7.00	8.89	10.92	15.75	28.00
8	0.88	1.60	2.48	3.52	4.80	6.32	8.00	10.16	12.48	18.00	32.00
9	0.99	1.80	2.79	3.96	5.40	7.11	9.00	11.43	14.04	20.25	36.00
10	1.10	2.00	3.10	4.40	6.00	7.90	10.00	12.70	15.60	22.50	40.00

TABLE C.2 Cross-Sectional Area (mm²) of Combinations of Metric Bars of the Same Size

Number of Bars	Bar Number							
	10	15	20	25	30	35	45	55
1	100	200	300	500	700	1 000	1 500	2 500
2	200	400	600	1 000	1 400	2 000	3 000	5 000
3	300	600	900	1 500	2 100	3 000	4 500	7 500
4	400	800	1 200	2 000	2 800	4 000	6 000	10 000
5	500	1 000	1 500	2 500	3 500	5 000	7 500	12 500
6	600	1 200	1 800	3 000	4 200	6 000	9 000	15 000
7	700	1 400	2 100	3 500	4 900	7 000	10 500	17 500
8	800	1 600	2 400	4 000	5 600	8 000	12 000	20 000
9	900	1 800	2 700	4 500	6 300	9 000	13 500	22 500
10	1 000	2 000	3 000	5 000	7 000	10 000	15 000	25 000

TABLE C.3 Cross-sectional Area per Foot Width (in.²/ft) of Bars of the Same Size

Bar Spacing (in.)	Bar Number										
	3	4	5	6	7	8	9	10	11	14	18
4	0.33	0.60	0.93	1.32	1.80	2.37	3.00	3.81	4.68		
4½	0.29	0.53	0.83	1.17	1.60	2.11	2.67	3.39	4.16	6.00	
5	0.26	0.48	0.74	1.06	1.44	1.90	2.40	3.05	3.74	5.40	9.60
5½	0.24	0.44	0.68	0.96	1.31	1.72	2.18	2.77	3.40	4.91	8.73
6	0.22	0.40	0.62	0.88	1.20	1.58	2.00	2.54	3.12	4.50	8.00
6½	0.20	0.37	0.57	0.81	1.11	1.46	1.85	2.34	2.88	4.15	7.38
7	0.19	0.34	0.53	0.75	1.03	1.35	1.71	2.18	2.67	3.86	6.86
7½	0.18	0.32	0.50	0.70	0.96	1.26	1.60	2.03	2.50	3.60	6.40
8	0.17	0.30	0.47	0.66	0.90	1.19	1.50	1.91	2.34	3.38	6.00
8½	0.16	0.28	0.44	0.62	0.85	1.12	1.41	1.79	2.20	3.18	5.65
9	0.15	0.27	0.41	0.59	0.80	1.05	1.33	1.69	2.08	3.00	5.33
9½	0.14	0.25	0.39	0.56	0.76	1.00	1.26	1.60	1.97	2.84	5.05
10	0.13	0.24	0.37	0.53	0.72	0.95	1.20	1.52	1.87	2.70	4.80
10½	0.13	0.23	0.35	0.50	0.69	0.90	1.14	1.45	1.78	2.57	4.57
11	0.12	0.22	0.34	0.48	0.65	0.86	1.09	1.39	1.70	2.45	4.36
11½	0.11	0.21	0.32	0.46	0.63	0.82	1.04	1.33	1.63	2.35	4.17
12	0.11	0.20	0.31	0.44	0.60	0.79	1.00	1.27	1.56	2.25	4.00
13	0.10	0.18	0.29	0.41	0.55	0.73	0.92	1.17	1.44	2.08	3.69
14	0.09	0.17	0.27	0.38	0.51	0.68	0.86	1.09	1.34	1.93	3.43
15	0.09	0.16	0.25	0.35	0.48	0.63	0.80	1.02	1.25	1.80	3.20
16	0.08	0.15	0.23	0.33	0.45	0.59	0.75	0.95	1.17	1.69	3.00
17	0.08	0.14	0.22	0.31	0.42	0.56	0.71	0.90	1.10	1.59	2.82
18	0.07	0.13	0.21	0.29	0.40	0.53	0.67	0.85	1.04	1.50	2.67

TABLE C.4 Cross-sectional Area per Meter Width (mm²/m) of Metric Bars of the Same Size

Bar Spacing (mm)	Bar Number									
	10	15	20	25	30	35	45	55
100	1 000	2 000	3 000	5 000	7 000	10 000	15 000	20 000
125	800	1 600	2 400	4 000	5 600	8 000	12 000	16 667
150	667	1 333	2 000	3 333	4 667	6 667	10 000	14 286
175	571	1 143	1 714	2 857	4 000	5 714	8 571	12 500
200	500	1 000	1 500	2 500	3 500	5 000	7 500	10 000
250	400	800	1 200	2 000	2 800	4 000	6 000	8 333
300	333	667	1 000	1 667	2 333	3 333	5 000	7 143
350	286	571	857	1 429	2 000	2 857	4 286	6 250
400	250	500	750	1 250	1 750	2 500	3 750	5 555
450	222	444	666	1 111	1 555	2 222	3 333	5 555

Appendix D

Conversion Between Conventional American and SI Metric Units

CONVERSION FROM CONVENTIONAL AMERICAN TO METRIC UNITS

Length

1 ft	$= 0.304\ 8$ m $= 304.8$ mm
1 in.	$= 25.40$ mm

Area

1 ft^2	$= 0.092\ 903$ m^2 $= 92\ 903$ mm^2
1 in.^{2}	$= 645.16$ mm^2

Volume, Section Modulus

1 ft^3	$= 0.028\ 316$ m^3 $= 28\ 316\ 000$ mm^3
1 in.^{3}	$= 16\ 387$ mm^3

Moment of Inertia

1 ft^4	$= 0.008\ 631$ m^4 $= 8\ 631\ 000\ 000$ mm^4
1 in.^{4}	$= 416\ 231$ mm^4

Mass

1 lb	$= 0.453\ 59$ kg

Weight

1 k	= 1000 lb = 4.448 22 kN
1 lb	= 4.448 22 N

Force

1 k	= 1000 lb = 4.448 22 kN
1 lb	= 4.448 22 N

Force per Unit Length

1 klf	= 14.594 kN/m
1 plf	= 14.594 N/m

Force per Unit Area

1 psf	= 47.880 Pa
1 psi	= 6.894 76 kPa

Moment

1 k-ft	= 1.355 82 kN·m
1 lb-ft	= 1.355 82 N·m
1 k-in.	= 112.985 N·m
1 lb-in.	= 0.112 985 N·m

Moment per Unit Width

1 k-ft per foot	= 4.448 22 kN·m/m
1 lb-ft per foot	= 4.448 22 N·m/m

Stress, Pressure

1 psf	= 47.880 Pa
1 ksi	= 1,000 psi = 6.894 76 MPa
1 psi	= 6.894 76 kPa

CONVERSION FROM METRIC TO CONVENTIONAL AMERICAN UNITS

Length

1 m	= 3.280 833 ft = 39.370 in.
1 mm	= 0.039 370 in.

Area

1 m^2	= 10.764 ft^2 = 1,550.003 in.^{2}
1 mm^2	= 0.001 550 in.^{2}

Volume, Section Modulus

1 m^3	= 35.315 ft^3
1 mm^3	= 0.000 061 024 in.^{3}

Moment of Inertia

1 m^4	= 2,402,500 in.^{4}
1 mm^4	= 0.000 002 403 in.^{4}

Mass

1 kg	= 2.204 62 lb

Weight

1 MN	= 224.809 k = 224,809 lb
1 kN	= 224.809 lb

Force

1 MN	= 224,809 lb
1 kN	= 224.809 lb

Force per Unit Length

1 MN/m	= 68.522 klf = 68,522 plf
1 kN/m	= 68.522 plf

Force per Unit Area

1 MPa	= 20.885 434 ksf = 145.037 738 psi
1 kPa	= 20.885 434 psf = 0.145 038 psi

Moment

1 MN·m	= 737.562 k·ft
1 kN·m	= 737.562 lb·ft = 8.850 732 k·in.

Moment per Unit Width

1 MN·m/m	= 224.809 k·ft per foot
1 kN·m/m	= 224.809 lb·ft per foot

Stress, Pressure

1 MPa	= 20.885 434 ksf = 145.037 738 psi
1 kPa	= 20.885 434 psf = 0.145 038 psi

Appendix E

Answers to the Numerical Questions

CHAPTER 4

2. **(a)** (i) 69.190 m; (ii) 3.048 m; (iii) 1.753 m; (iv) 203 mm; (v) 114 mm; (vi) 19 mm; (vii) 13 mm

(b) (i) 272 ft 4 in.; (ii) 9 ft 10 in.; (iii) 5 ft 10 in.; (iv) 1 ft 0 in.; (v) 4½ in.; (vi) ½ in.; (vii) ⅜ in.

(c) 318 mm²/m

(d) 0.14 in.²/ft

(e) 339 kN·m

(f) 282 kN·m

(g) 295,000 lb·ft = 3,540,000 lb·in.

(h) 22.24 kN·m/m

(j) 4,496 lb·ft/ft

(k) 9.58 kPa

(m) 209 psf

(n) 27.6 MPa and 414 MPa

(p) 4350 psi and 58,000 psi

CHAPTER 6

2. **(a)** ¾ in. (provided reinforcing bars are No. 11 or smaller)
   (b) 3 in.

(c) $1\frac{1}{2}$ in. if concrete exposed to weather; $\frac{3}{4}$ in. if protected from weather (provided bars are below No. 6 in the first case, and below No. 14 in the second case)

(d) $1\frac{1}{2}$ in.

(e) 2 in. (provided main reinforcing bars are No. 6 or larger)

3. For Grade 60 reinforcement: 0.18% for all reinforcement; 0.34% for positive reinforcement in beams

4. For beams: 1 in. or the bar diameter, whichever is the bigger. For columns: $1\frac{1}{2}$ in. or $1\frac{1}{2}$ times the bar diameter, whichever is the bigger

5. 18 in. or 3 times the wall or slab thickness, whichever is smaller

CHAPTER 7

5. 29,000,000 psi

6. 3,120,000 psi

7. 60 psf

8. 3 kPa

CHAPTER 8

3. 200%

CHAPTER 9

9. $w_u = 1.4 \times 90 + 1.7 \times 60 = 228$ psf
$M_u = w_u \ell_n^2/10 = 228 \times 12.5^2/10 = 3{,}562.5$ lb·ft/ft $= 42{,}750$ lb·in./ft

10. $h = \ell/24 = 12.5 \times 12/24 = 6.25$ in.   *say $6\frac{1}{2}$ in.*

11. *Assuming No. 4 bars*, $d = 6.5 - \frac{3}{4} - \frac{1}{4} = 5.5$ in.
$z = 0.95 \times 5.5 = 5.23$ in.
$A_s = 42{,}750/0.9 \times 60{,}000 \times 5.23 = 0.1514$ in.^{2}/ft   *main reinforcement*
Minimum reinforcement $= 0.0018 \times 12 \times 6.5 = 0.1404$ in.^{2}/ft   *secondary reinforcement*

12. Maximum permissible spacing 18 in. or $3 \times 6\frac{1}{2} = 19.5$ in. *Use No. 4 at 15 in. as main reinforcement, and No. 4 at 16 in. as secondary reinforcement.*

13. $z/d = 0.94$
$6000 = 0.9 \times 60{,}000 \times 0.006 \times 0.94 \times 1 \times d^2$ (in inches); $d = 4.44$ in.
$A_s = 0.006 \times 12 \times 4.44 = 0.320$ in.^{2}/ft

14. *No. 4 at 7½ in. centers*; $h = 5½$ *in.*

15. $z/d = 0.95$
    $80,000 = 0.9 \times 40,000 \times 0.007 \times 0.95 \times 12 \times d^2$
    $d = 5.28$ in.
    $A_s = 0.007 \times 12 \times 5.28 = 0.444$ in.^{2}/ft

16. *No. 5 at 8 in.*; $h = 6½$ *in.*

17. $d = 5.5$ in.; $A_s = 0.23$; $\rho = 0.23 / 12 \times 5.5 = 0.0035$
    $z/d = 0.95$; $z = 5.23$ in.
    $M' = 0.9 \times 60,000 \times 0.23 \times 5.23 = 64,957$ lb·in./ft $= 5413$ lb·ft/ft

18. $M_u = w_u \ell^2/8 = 5413 = w_u \times 12.5^2 / 8$
    $w_u = 277$ psf
    $1.4 \times 90 = 126$; $1.7L = 277 - 126 = 151$
    $L = 88.82$, *88 psf*

19. $d = 7$ in.; $A_s = 0.34$; $\rho = 0.0040$; $z/d = 0.96$; $z = 6.72$ in.
    $M' = 0.9 \times 60,000 \times 0.34 \times 6.72 = 123,379$ lb·in./ft $=$
    $10,282$ lb·ft/ft

20. M_{max} (at first interior support) $= 0.105 \times 1.4D \, \ell^2 + 0.116 \times 1.7L \ell^2$
    $= 0.105 \times 1.4 \times 900 \times 32^2 + 0.116 \times 1.7 \times 550 \times 32^2 =$
    $246,538$ lb·ft $= 2,958,456$ lb·in.

21. *Take* $\rho = 0.18f'_c/f_y = 0.18 \times 4000 / 60,000 = 0.012$; $z/d = 0.89$
    $bd^2 = 2,958,456/0.9 \times 60,000 \times 0.012 \times 0.89 = 5130$ in.^{3}.
    If $d = 2b$, $b^3 = 1283$ and $b = 10.86$ in. *Taking b = 12 in.*,
    $d = 20.68$ in.
    $A_s = 0.012 \times 12 \times 20.68 = 2.978$ in.^{2}

22. *Choose 4 No. 8*; $h = 20.68 + 0.50 + 0.38 + 1.50 = 23.06$ in.
    24 in.

23. *Take* $\rho = 0.012$ *(see Exercise 21)*; $z/d = 0.89$
    $bd^2 = 300,000 \times 12 / 0.9 \times 60,000 \times 0.012 \times 0.89 = 6242$ in.^{3}.
    If $d = 2b$, $b^3 = 1561$, $b = 11.60$, *choose b = 12 in.* $d = 22.81$ in.
    $A_s = 0.012 \times 12 \times 22.81 = 3.28$ in.^{2}.

24. $A_s = 4.80$ in.^{2}
    $d = 30 - 1.50 - 0.38 - 0.88 - 0.50 = 26.74$ in.
    $\rho = 4.80 / 15 \times 26.74 = 0.0120$
    $z/d = 0.87$; $z = 23.26$ in.
    $M' = 0.9 \times 60,000 \times 4.80 \times 23.26 = 6,028,992$ lb·in. $= 502,416$ lb·ft

25. $w_u \ell^2/8 = 502,416$ lb·ft; $\ell = 32$ ft; $w_u = 3925$ lb/ft
    $1.4D = 1680$ lb/ft; $1.7L = 3925 - 1680 = 2245$ lb/ft; $L = 1321$ lb/ft

26. $A_s = 3.60$ in.^{2}
    $d - 30 - 1.50 - 0.38 \quad 0.88 \quad 0.50 = 26.74$ in.
    $\rho = 3.60 / 12 \times 26.74 = 0.0112$
    $z/d = 0.92$; $z = 24.60$ in.
    $M' = 0.9 \times 60,000 \times 3.60 \times 24.60 = 4,782,240$ lb·in. $= 398,520$ lb·ft

Metric Exercises

27. $w_u = 1.4 \times 4.5 + 1.7 \times 2.8 = 11.06$ kPa
$M_u = 11.06 \times 3.75^2 / 10 = 15.553$ kN·m/m

28. $h = 3750/24 = 156$ mm   *say 175 mm*

29. *Assuming No. 10 bars*: $d = 175 - 20 - 5 = 150$ mm
$z = 0.95 \times 150 = 143$ mm
$A_s = 15.553 \times 10^{-3} / 0.9 \times 400 \times 0.143 = 302 \times 10^{-6}$ m²/m $=$
302 mm²/m   *main reinforcement*
minimum reinforcement $= 0.0018 \times 1000 \times 150 =$
270 mm²/m   *secondary reinforcement*

30. Maximum permissible spacing $= 450$ mm or $3 \times 175 = 525$ mm
use No. 10 at 300 mm and at 350 mm

31. $z/d = 0.95$
$24 \times 10^{-3} = 0.9 \times 400 \times 0.006 \times 0.95 \times 1 \times d^2$ (in m);
$d = 108$ mm
$A_s = 0.006 \times 1000 \times 108 = 648$ mm²/m

32. *No. 15 at 300 mm*; $h = 108 + 8 + 20 = 136$ *say 150 mm*

33. $d = 150 - 20 - 8 = 122$ mm
$A_s = 1000$ mm²/m; $\rho = 1000/1000 \times 122 = 0.008\ 197$
$z/d = 0.92$; $z = 112$ mm
$M' = 0.9 \times 400 \times 1000 \times 10^{-6} \times 0.112$ (in MN·m/m) $=$
40.32 kN·m/m

34. $M_u = w_u \ell^2/8 = 40.32 = w_u \times 4.5^2 / 8$; $w_u = 15.92$ kPa
$1.4 \times 4 = 5.60$ kPa;
$1.7L = 15.92 - 5.60$; $L = 6.07$ kPa

35. $A_s = 1500$ mm²/m; $d = 400 - 20 - 10 = 370$ mm;
$\rho = 1500/1000 \times 370 = 0.004\ 054$
$z/d = 0.97$; $z = 359$ mm
$M' = 0.9 \times 400 \times 1500 \times 10^{-6} \times 0.359$ (in MN/m) $= 193.86$ kN/m

36. M_{max} (at first interior support) $= 0.105 \times 1.4D\ell^2 + 0.116 \times 1.7L\ell^2$
$= 0.105 \times 1.4 \times 10 \times 10.5^2 + 0.116 \times 1.7 \times 7.5 \times 10.5^2 =$
325.13 kN·m

37. *Take* $\rho = 0.18\ f'_c/f_y = 0.18 \times 25/400 = 0.011$; $z/d = 0.89$
$bd^2 = 325.13 \times 10^{-3} / 0.9 \times 400 \times 0.011 \times 0.89 =$
92.25×10^{-3} m³.
If $d = 2b$, $b^3 = 0.023\ 063$ m³ and $b = 0.284$ m.
Taking b = 300 mm, d = 555 mm
$A_s = 0.011 \times 300 \times 555 = 1832$ mm².

38. *4 No. 25 bars*; $h = 555 + 13 + 10 + 40 = 618$ mm   *700 mm*

39. *Take* $\rho = 0.18 \times 20/400 = 0.009$; $z/d = 0.89$
$bd^2 = 400 \times 10^{-3} / 0.9 \times 400 \times 0.009 \times 0.89 = 138.72 \times 10^{-3}$ m³.
For $d = 2b$, $b = 326$ mm. *Take b = 400 mm*, gives $d = 589$ mm
$A_s = 0.009 \times 400 \times 589 = 2120$ mm².

40. $A_s = 2400$ mm^2
   $d = 750 - 40 - 10 - 20 - 13 = 667$ mm
   $\rho = 2400 / 375 \times 667 = 0.0096$
   $z/d = 0.89; z = 594$ mm
   $M' = 0.9 \times 400 \times 2400 \times 10^{-6} \times 0.594$ (MN·m) $= 513.2$ kN·m
41. $w_u = 513.2 \times 8 / 9^2 = 50.69$ kN/m
   $1.4 \times 16 = 22.40$ kN/m
   $L = (50.69 - 22.4) / 1.7 = 16.6$ kN/m
42. $A_s = 1800$ mm^2
   $d = 750 - 40 - 10 - 20 - 13 = 667$ mm
   $\rho = 1800 / 300 \times 667 = 0.0090$
   $z/d = 0.93; z = 620$ mm
   $M' = 0.9 \times 400 \times 1800 \times 10^{-6} \times 0.620$ (MN·m) $= 401.8$ kN·m

CHAPTER 10

10. *Data given*: $b = 18$ in., $b_w = 10$ in., $h_f = 3$ in., $d = 20$ in.,
   $M_u = 3,000,000$ lb·ft, $f'_c = 3500$ psi, $f_y = 60,000$ psi, and $\phi = 0.9$.
   The ratio $h_f/d = 3/20 = 0.150$.
   From Eq. (10.4), $1.71 \times 3,000,000 / 3500 \times 18 \times 20^2 =$
   $0.204 > h_f/d$.
   The beam is therefore a T-beam and, from Eq. (10.3),
   $A_s = 3,000,000 / 0.9 \times 60,000(20 - \frac{1}{2} \times 3) = 3.00$ in.^{2}
11. *Data given*: $b = 10$ in., $d = 18$ in., $M_u = 3,000,000$ lb·ft,
   $f'_c = 3500$ psi, $f_y = 60,000$ psi, $\phi = 0.9$, $k_m = 772$ MPa, and
   $\rho_{max} = 0.0187$.
   From Eq. (10.5), $M_1 = 772 \times 10 \times 18^2 = 2,501,280$ lb·ft.
   This is less than M_u, so compression reinforcement is required.
   $M_2 = 3,000,000 - 2,501,280 = 498,720$ lb·ft.
   Assuming one row of No. 7 bars for the compression reinforcement and
   No. 3 stirrups, $d' = 1.5 + 0.38 + \frac{1}{2} \times 0.88 = 2.32$ in.,
   and $d - d' = 15.68$ in.
   $A_{s1} = 0.0187 \times 10 \times 18 = 3.366$ in.^{2}; $A_{s2} = 498,720 / 60,000 \times 15.68$
   $= 0.530$ in.^{2};
   $A_s = 3.366 + 0.530 = 3.896$ in.^{2} and $A'_s = 0.530$ in.^{2}

Metric Exercises

12. *Data given*: $b = 450$ mm, $b_w = 250$ mm, $h_f = 75$ mm, $d = 500$ mm,
   $M_u = 400$ kN·m, $f'_c = 20$ MPa, $f_y = 400$ MPa, and $\phi = 0.9$.
   The ratio $h_f/d = 75/500 = 0.150$.
   From Eq. (10.4), $1.71 \times 400 \times 10^{-3}/20 \times 0.45 \times 0.5^2 -$
   $0.304 > h_f/d$.
   The beam is therefore a T-beam, and from Eq. (10.3),
   $A_s = 400 \times 10^{-3} / 0.9 \times 400(0.5 - \frac{1}{2} \times 0.075) = 2.402 \times 10^{-3}$ m^2
   $= 2402$ mm^2.

13. *Data given*: $b = 250$ mm, $d = 450$ mm, $M_u = 300$ kN·m, $f'_c = 20$ MPa, $f_y = 400$ MPa, $\phi = 0.9$, $k_m = 4.73$ MPa, and $\rho_{max} = 0.0162$.
   From Eq. (10.5), $M_1 = 4.73 \times 0.25 \times 0.45^2 = 239.46 \times 10^{-3}$ MN·m.
   This is less than M_u, so compression reinforcement is required.
   $M_2 = 300 - 239.46 = 60.54$ kN·m.
   Assuming No. 20 bars in one layer for the compression reinforcement and No. 10 stirrups, $d' = 40 + 10 + \frac{1}{2} \times 20 = 60$ mm;
   $d - d' = 390$ mm;
   $A_{s1} = 0.0162 \times 250 \times 450 = 1823$ mm^2;
   $A_{s2} = 60.54 \times 10^{-3} / 400 \times 0.39 = 0.388 \times 10^{-3}m^2 = 388$ mm^2;
   $A_s = 2211$ mm^2; $A'_s = 388$ mm^2.

CHAPTER 11

17. The effective depth $d = 28 - 1.5 - 0.38 - \frac{1}{2} \times 1.13 = 25.55$ in.;
   $w_u = 1.4 \times 1200 + 1.7 \times 700 = 2870$ lb/ft;
   $V_u = \frac{1}{2} \times 2870 \times 26$ (a slight overestimate) $= 37,310$ lb;
   $V_n = 37,310 / 0.85 = 43,894$ lb;
   $v_n = 43,894/16 \times 25.55 = 107$ psi.
   Since $2\sqrt{f'_c} = 118$ psi, only nominal shear reinforcement is required.
   Maximum spacing is 24 in. or $\frac{1}{2}d = 12.7$ in.
   Maximum spacing of No. 3 stirrups (the smallest available, $A_v = 0.22$ in.^{2}) to provide the equivalent of an ultimate shear stress of 50 psi is $s = 60,000 \times 0.22 / 50 \times 16 = 16.5$ in. *Provide No. 3 stirrups at 12- in. centers*.

18. $w_u = 3380$ lb/ft; $V_u = 43,940$ lb; $V_n = 51,694$ lb; $v_n = 126$ psi; $2\sqrt{f'_c} + 50 = 168$ psi.
   Although the shear force is big enough to require reinforcement, the nominal reinforcement is still adequate. *Provide No. 3 stirrups at 12-in. centers*.

CHAPTER 12

11. The minimum thickness from Table 8.3 is $3\frac{1}{2}$ in.
   The clear span is $\ell_n = 18 \times 12 - 12 = 204$ in.
   The minimum thickness for deflection control, from Eq. (8.5), is
   $h = (204 / 36,000) (800 + 60,000 / 200) = 6.23$ in.
   Use a $6\frac{1}{2}$-in.-thick slab.

12. The minimum thickness from Table 8.3 is 5 in.
   The clear span, as in Exercise 11, is $18 \times 12 - 12 = 204$ in.
   The minimum thickness for deflection control is therefore, as in Exercise 11, 6.23 in., as the same equation applies.
   The minimum thickness by shear resistance about the column is obtained

from the shear force. To calculate that, we must estimate the effective depth, but a small error in that makes very little difference to the result. *Assume an 8-in. slab reinforced with No. 4 bars in both directions*, so the smaller effective depth is $d = 8 - 0.75 - 0.5 - 0.25 = 6.5$ in. The side of the square around the column is
$12 + 6.5 = 18.5$ in. $= 1.54$ ft.
The circumference $b_0 = 4 \times 18.5 = 74$ in.
$w_u = 1.4 \times 110 + 1.7 \times 60 = 256$ psf.
$V_u = 256(18^2 - 1.54^2) = 82,337$ lb, and $V_n = 96,867$ lb. Maximum permissible shear stress is $4 \sqrt{f_c'} = 237$ psi.
$d = 96,867 / 74 \times 237 = 5.52$ in.; $h = 5.52 + 0.25 + 0.50 + 0.75$
$\qquad\qquad\qquad\qquad\qquad\qquad = 7.02$ in. *Use a $7\frac{1}{2}$-in. slab.*

CHAPTER 13

10. *Data given*: $D = 1,500,000$ lb, $L = 600,000$ lb, $f_c' = 5000$ psi, $f_y = 60,000$ psi.
    $U = 1,500,000 \times 1.4 + 600,000 \times 1.7 = 3,120,000$ lb. *Design a square column with* $\rho_t = 0.025$ *approximately*.
    $3,120,000 = 0.56[0.85 \times 5000 (1 - 0.025) + 60,000 \times 0.025]A_g$;
    $A_g = h^2 = 987.19$ in.^{2}; $h = 31.4$ in.   *32-in.-square column.*
    $A_{st} = (5,571,000 - 0.85 \times 5000 \times 32^2) / (60,000 - 0.85 \times 5000) =$
    21.87 in.^{2} *12 No. 14 bars.*
    Use No. 4 ties at 24-in. centers.

11. *Data given*: $b = h = 24$ in., $A_{st} = 18.00$ in.^{2}, $f_c' = 4000$ psi, $f_y = 60,000$ psi. $A_g = 576$ in.^{2};
    $P_u = 0.56[0.85 \times 4000 \times 576 + (60,000 - 0.85 \times 4000) \times 18.00] =$
    $1,667,000$ lb.

Metric Exercises

12. *Data given*: $D = 6000$ kN, $L = 3000$ kN, $f_c' = 35$ MPa, $f_y = 400$ MPa.
    $U = 6000 \times 1.4 + 3000 \times 1.7 = 13,500$ kN $= 13.5$ MN. *Design a square column with* $\rho_t = 0.025$ *approximately*.
    $13.5 = 0.56[0.85 \times 35 (1 - 0.025) + 400 \times 0.025]A_g$;
    $A_g = h^2 = 0.6180$ m^2;
    $h = 786$ mm   *800-mm-square column.*
    $A_{st} = (24.11 - 0.85 \times 35 \times 0.8^2) / (400 - 0.85 \times 35) =$
    $0.013\ 693$ m$^2 = 13,693$ mm^2 *12 No. 45 bars.*
    Use No. 15 ties at 600-mm centers.

13. *Data given*: $b = h = 600$ mm, $A_{st} = 10,000$ mm^2, $f_c' = 35$ MPa, $f_y = 400$ MPa. $A_g = 0.36$ m^2;
    $P_u = 0.56[0.85 \times 35 \times 0.36 + (400 - 0.85 \times 35) \times 0.01] =$
    8.07 MN $= 8070$ kN.

CHAPTER 14

6. *For the purpose of calculating the shear around the column, assume an effective depth d = 24 in.* Thus the length of the critical line for shear is: column width + 2 × ½d = 24 + 24 = 48 in. = 4 ft.
The circumference of this line around the column is b_0 = 4 × 48 = 192 in. = 16 ft.
The shear force around the column is V_u = 4500 (14^2 − 4^2) = 810,000 lb; V_n = 952,941 lb.
The maximum permissible shear stress is $4\sqrt{f_c'}$ = 237 psi.
The minimum effective depth required is d = 952,941 / 192 × 237 = 20.94 in.

7. The span of the cantilever is ½ (14 − 2) = 6 ft.
M_u = ½ × 4500 × 6^2 = 81,000 lb·ft/ft.
Take z conservatively as 0.9d = 0.9 × 20.94 = 18.85 in.
A_s = 81,000 × 12 / 0.9 × 60,000 × 18.85 = 0.955 in.^{2}/ft. *Use No. 7 at 7½-in. centers, two layers at right angles to one another.*
h = 20.94 + ½ × 0.88 + 0.88 + 3.00 = 25.26 in.   *say 27 in.*

Index

LIBRARY
ST. LOUIS COMMUNITY COLLEGE
AT FLORISSANT VALLEY